QUANTUM PHYSICS FOR BEGINNERS & KIDS

Box Set for Curious Minds to Understand the Subatomic
World & Basic Concepts from Wave Theory to the
Uncertainty Principle

ALISHA KAPANI

Contents

Quantum Physics for Beginners Simplified

Quantum Physics for Kids

Quantum Physics for Beginners Simplified

UNDERSTAND THE SUBATOMIC WORLD,
APPLY BASIC CONCEPTS TO EVERYDAY LIFE,
AND EXPAND YOUR CONSCIOUSNESS &
WORLDVIEW WITHOUT A SCIENCE
BACKGROUND

Preface

My undergraduate education began in chemical engineering at the dynamic University of Calgary. However, not long after becoming a working professional engineer, my journey took an unexpected turn when I delved into the realms of quantum physics and quantum healing for my postgraduate studies at Quantum University in Hawaii. This exploration wasn't simply out of curiosity, however—it was a deeply personal quest for healing.

The core reason I wound up researching the world of quantum healing was my profound struggle with severe depression and various physical issues that lasted for years. Faced with these challenges and no workable solutions in sight, I turned to more unconventional paths to wellness, which led me to the fascinating domain of quantum physics. Rather than relying solely on traditional methods, I also leaned into the idea of expanding consciousness as a means to heal myself. As I found out more about the intricacies of quantum mechanics and their connection to healing, a passion was ignited within me to understand and harness the potential of quantum principles for healing. My approach has always been grounded in a genuine curiosity that pushes the boundaries of conventional wisdom; I have questioned everything in my pursuit of knowledge, including the limitations of what I was taught in classical physics and chemistry as an engineer.

Nowadays, I extend my knowledge and assistance to clients globally. Quantum healing has become my calling, a method through which I help others navigate any issues they face with their physical and mental well-being, just as I did my own.

One thing I have observed at my vocation, and the reason that I wrote this book, is that quantum physics is not as accessible to everyone as I would prefer it to be. I had the advantage of my engineering background and natural curiosity, and even I struggled at the beginning to comprehend these complex concepts. As such, I feel that there is an unmet need for simplicity when it comes to learning quantum physics.

This book serves as such a guide, with explanations and illustrative examples of the various concepts of quantum physics that do not require you to understand complex formulas or mathematics; it seeks to ensure that even non-engineers and people without a background in the sciences can embark on the same kind of explorative journey that I did. It also includes a section that gives a taste of my own field of expertise: quantum healing.

Through this book, I hope to leave this one message at the very least: the possibilities of the quantum world extend far beyond the limitations given by conventional education. If I had not explored beyond what I was taught, I may never have healed myself and become the person I am today. Quantum physics has the potential to revolutionize our understanding of the world, opening our minds to new dimensions of thought and healing.

Personally, the shift in perspective I experienced from my research has made me more open, curious, and receptive to the myriad opportunities that life presents. I am in favor of an open-minded approach to life, and I encourage my clients and others to explore all possibilities. My personal recovery from my health issues is a testament to the transformative power of embracing unconventional paths. By helping you understand the science behind quantum principles, I hope you may also become more receptive to the diverse array of opportunities it provides and, in the process, open your mind to newer and better ways to foster your personal growth and well-being!

Introduction

The Ultraviolet Catastrophe

Quantum Physics. The name itself often evokes a sense of something beyond our daily lives, something that, perhaps, can only be found and experienced in the depths of black holes or from interactions at a scale so tiny that we can never hope to observe them with the naked eye. What even is a quantum, anyway?

That question will be answered very shortly! But first, the common misconception above needs to be dispelled. Even though we may not be able to see much of quantum physics in action without the aid of highly advanced equipment and technology, its effects are very much a part of our day-to-day life.

Consider the toaster you might use to make your breakfast some mornings. Or the coals underneath a barbeque grill on a fine summer evening. In both cases, something that looks fairly dark and grayish at room temperature starts to glow red when heated:

the heating element in the toaster and the coals themselves in the grill. The latter starts to glow orange and even yellow when heated further.

This glow, or emission of light, does not depend on the material being heated—both a piece of glass and a piece of iron will glow the same shades of red, orange, and yellow if they can be heated to the same temperatures. That's because the color of this glow (specifically, the wavelength of the light it is giving off) is dependent on just that: the temperature of an object. This is one of the key underpinnings of the phenomenon known as "black-body radiation": the radiation of light and other energies that are given off by an object due to its temperature. While the definition of a "black body" isn't important at this point, it's important to note that light is what's called "electromagnetic radiation," and the rest of the energies given off by a hot object are also electromagnetic by nature (such as infrared waves).

One of the earliest attempts to derive the relationship between the electromagnetic radiation emitted by hot objects and their temperature was the Rayleigh-Jeans Law, proposed by physicists Lord Rayleigh and Sir James Jeans, which was based on experimental findings and classical wave theories in physics at the beginning of the 20th century. Their law treated hot objects a little like a guitar: The same way that playing all the strings on a guitar releases sound from each of them at equal volume, a hot object would emit radiation of equal energy at all the different wavelengths that it can do so for a given temperature.

However, the theory ran into a problem: While it correctly predicted the intensities of the electromagnetic radiation given off at higher wavelengths—such as microwaves, infrared waves, and light—it also predicted that any hot object emitting radiation would give off near-infinite levels of low-wavelength radiation, such as ultraviolet waves, X-rays, and radioactive gamma rays. This is due to how, in classical wave theory, energy has a continuous and inverse relationship with wavelength in a black body. As such, the shorter the wavelength, the greater the energy released, according to the theory.

You know, from your daily experience, that this is not the case, though—your slices of bread are not bombarded by an atom bomb's worth of nuclear radiation every time you toast them! However, the classical theories of electromagnetic waves at the time could not explain why this was the case. Since the theory started to noticeably part ways with reality when dealing with ultraviolet radiation emissions, this failure of classical physics came to be known as the "Ultraviolet Catastrophe."

The solution to the catastrophe started to take shape when a physicist named Max Planck suggested something quite radical at the time: That electromagnetic energy was not released continuously as waves by hot objects but instead in discrete quantities that were determined by their characteristic wavelength. In simple terms, rather than being like ketchup pouring out continuously from a bottle, the energy was instead more like ketchup packets given out by a dispenser. These packets of energy were, in keeping with a recurring scientific tradition, named after the Latin analog for the packet, or discrete amount, "quantum."

Since the amount of energy in a quantum was inversely dependent on its associated wavelength, this meant that at lower wavelengths, there was a naturally induced cap on how many such packets of energy could be given out by a hot object. This cap then became lower as the wavelength grew smaller—after all, an object cannot give out more energy than it possesses! The treatment of energy as discrete packets, or the quantization of energy, led to a theory more closely aligned with what was experimentally observed.

This theory of the quantization of energy was the foundation on which the field of quantum physics, the branch of physics that mainly deals with energies at a very, very small scale, was built. And, owing to some further quirks that would be discovered in the course of the 20th and 21st centuries, quantum physics also deals with the behavior of matter at a very minute scale.

The Uncertain Potential of Quantum Physics

The idea of energy behaving like packets as opposed to a continuous flow was revolutionary due to one major reason: it introduced an alternative perspective to physics from the classical approach that had dominated science for centuries. As was the case for black-body radiation, it was able to explain multiple odd observed behaviors that classical physics could not. However, quantum physics wasn't quite done destabilizing the accepted understanding of physics at the beginning of the 20th century. It also poked a hole in another foundational belief regarding physics and science in general—that of determinism.

Determinism is the idea that we can predict how something will behave in the future based on the knowledge of certain initial conditions. For example, if you throw a football up in the air, you not only know that it will eventually fall back down again, but you can also know how much time it would take to come back down and where it will land. Essentially, you are predicting how the ball will behave based on the initial conditions in which you threw it, and classical physics is what allows you to do this.

However, this deterministic worldview breaks down at the quantum level, and it becomes very difficult—and, in some aspects, impossible—to predict how particles and energies will behave depending on their initial conditions. In fact, even observing those initial conditions becomes an exercise in probability!

For example, one of the key concepts of quantum physics is the principle of superposition. You may be more familiar with the thought experiment that illustrates it, though. Schrödinger's Cat is named after the physicist who came up with it, Erwin Schrödinger. According to classical physics, the cat will be alive in the box until it dies, after which point it will be dead. But if the cat was a quantum cat, it would be simultaneously alive and dead at all times and only become either of the two when the box is opened and the cat is observed (Garisto, 2022).

This idea illustrates the peculiarities of certain properties at the quantum level, such as the position of an electron in an atom. While the classical view of an electron sees it as orbiting around the middle of the atom, what has been observed instead is that the electron's many possible positions in the atom are actually superimposed onto each other. It is only when the electron's position is directly observed that it "collapses" into a single point, much like how the cat is determined to be alive or dead when observed. So, until it is observed, the electron could be anywhere in the space it occupies in the atom—its position is uncertain. And this is merely a fraction of the uncertainty and indeterminism that quantum physics has in store.

It is precisely because of all these peculiarities and irregularities that quantum physics has not only changed the way we understand the world at a microscopic level but also led the way to groundbreaking technological advancements, such as semiconductors and quantum computers. The implications of quantum physics go far beyond laboratories and technology; it's a key part of unraveling the fabric of the universe itself. Quantum physics can help us even understand the way that black holes, stars, galaxies, and the cosmos itself behave.

All these peculiarities, implications, and more will be covered in depth in the chapters to come. Each chapter will be dedicated to a singular, fundamental concept of quantum physics, from wave-particle duality to quantum entanglement. These are concepts that take a fairly in-depth understanding of math and physics to truly grasp, but they will be introduced and explained with a minimum of required mathematical knowledge in the course of this book.

ONE

Introduction to Quantum Physics

The Demise of Local Realism

I n 2022, the Nobel Prize for Physics was awarded to three physicists who proved something rather disturbing: that the universe is not real. At least not in the way that we think it is.

The three scientists' names are John Clauser, Alain Aspect, and Anton Zeilinger. What they specifically proved through decades of experimentation, which began in the '70s, was that the universe is not locally real. To understand the ramifications of this discovery, let's break down what "locally real" means.

"Real," in this sense, means that something has properties that stay the same no matter how they are observed or what they're interacting with. For example, a blue ball will be blue whether you look at it, throw it, or ignore it. "Local" here means that anything can only affect and be affected by other objects or effects in their vicinity. These external influences can't travel faster than the speed of light, so to throw the ball, you need to hold it and exert force on it at close range. Even the color of the ball is caused by particles of light, called photons, bouncing off its surface before entering your eyes—and these interactions also take place in the vicinity of the ball.

What the three physicists showed with their research was that these aspects of a thing cannot both be true at the same time. This means that objects may not have definitive properties before being observed or interacted with, and they could also be influenced by very distant influences. Going back to the ball, the implications are that it may not always be blue depending on how it is observed, and it could potentially be disturbed by an entity that is *millions* of miles away.

You can see why such a result might be both very impactful on our understanding of the universe and extremely worrisome. There is a popular philosophical question: "If a tree falls in a forest and no one is around to hear it, does it make a sound?" For the longest time, what we understood about physics suggested that the tree would still make a sound even with no one around. However, in a universe that is not locally real, there is a chance the tree doesn't even exist unless someone observes it.

These are the kinds of wrenches that the field of quantum mechanics has been throwing into our understanding of how things work ever since its emergence in the last century. Even notable physicists, like the infamous Albert Einstein, have found it to be confounding: he once exasperatedly asked a colleague, "Do you really believe the moon is not there when you are not looking at it?"

Quantum Fundamentals

The field of quantum science deals with things that are very, very small in a scale of existence that is far removed from balls, trees, and moons. The observations made from this field are also on the same scale, meaning that there is no immediate danger of the moon vanishing from existence without an observer. It also means that to understand the inner workings of quantum physics, it's important to be placed in the right frame of perspective: the world of particles and photons in which quantum mechanics hold sway.

Everything that exists in the universe can be roughly divided into two categories: *energy* and *matter*. Energy refers to things like heat, light, sound, and electricity. They all have in common, and they cause changes to occur when objects interact with one another. In physical terms, energy is the property that is transferred during interactions. For example, when you boil an egg, heat is transferred into the egg when the boiling water interacts with it. When you kick a football, energy from your leg is transferred to the ball in the form of kinetic energy, the energy that moving objects have.

Matter is the stuff that most objects are made of. The air you breathe, the water you drink, and even your own body—these are all made up of matter. Matter comes in many different forms or states: the gaseous state that air is in, the liquid state of room-temperature water, and the solid state of your flesh and bones. Things can often be a mix of the different states as well. For example, carbonated water consists of bubbles made from carbon dioxide gas mixed with liquid water.

The Ancient Greek scholar Democritus was one of the first to theorize that all matter was made up of smaller, fundamental building blocks, which he called "atoms" at the time from the

Greek word for "indivisible" (since these were fundamental pieces, it was assumed they couldn't be broken down further). It took many centuries before technology was advanced enough to prove whether this was true or not.

As it understandably turned out, thinkers like Democritus were just off the mark: While the fundamental building blocks of matter, known as atoms, were proven to exist, they could be divided even further into smaller, more elementary particles. Research at the turn of the 20th century, such as J.J. Thompson's cathode-ray-tube experiment and Ernest Rutherford's alpha-particle-scattering experiment, helped build a picture of what the structure of an atom is.

An atom has two parts: a *nucleus* in its center, where most of its mass is, and a spherical space or "cloud" around it, which contains particles called *electrons*. The nucleus has two types of particles within it: *protons* and *neutrons*. To give you some idea of the proportions here, imagine that the whole atom is the size of a football stadium. The nucleus would be the size of a pea in the center of the field, and the rest of the stadium would be the cloud in which the atom's electrons exist.

These particles have two primary properties: *mass*, which determines how heavy they are or how much they are affected by gravity, and *electric charge*, which determines how they behave when in the presence of an electric field. While protons and neutrons are roughly the same in mass, electrons are much lighter than protons (one proton is equal in mass to roughly 1800 electrons). Protons have a positive electrical charge, electrons are negatively charged, and neutrons are, as their name implies, neutral in charge.

In the '60s, it was discovered that, while the electron was indeed an elementary particle, the proton and neutron were made up of even smaller particles. During this time, physicists were conducting experiments in order to discover and catalog all the various types of particles in existence, and their primary means of doing this was by smashing atoms together at very high speeds—in the same way that smashing an expensive clock can reveal the different parts it's made up of. While the latter is definitely not a recommended practice, the smashing of atoms together or bombarding them with energy causes interactions to take place that give rise to (or evidence for) new particles.

It's not easy to smash atoms together, though, especially when they need to be accelerated to very high speeds and collide under precise, heavily monitored conditions. Think about how much distance a plane needs to cover on the ground before it can move fast enough to take off or how much distance a car needs to drive to reach its top speed. Atoms need to reach much higher speeds than either a plane or a car before they reveal anything useful from being smashed, which is why the specialized buildings used for this purpose—usually called "particle accelerators" or "colliders"—are so large and often occupy many kilometers' worth of space.

When protons and neutrons were smashed, they were revealed to have an internal structure of what would eventually be called "quarks" by one of the physicists who theorized their existence, Murray Gell-Mann. *Quarks*, like electrons, are also elementary particles in that they cannot be divided into even smaller bits (at least, as far as we can tell with our current level of scientific knowledge and technology).

The study of not just these particles but many of the others that have been discovered through collider experiments, appropriately called *particle physics*, is a very closely related field to that of quantum physics. After all, they both study how reality operates at a microscopic scale. But, while particle physics is more concerned with the properties and categorization of both particles and the forces they exert on each other, quantum physics deals more with the behaviors of these particles and forces, which also includes how random and unpredictable these behaviors can be.

Photons and Atomic Energy Levels

In the previous chapter, how energy can be treated as discrete amounts or quanta (the plural of quantum) was discussed to help explain observations like the radiation given off by a heated object. The physicist who came up with the concept, Max Planck, approached it from a more mathematical than physical perspective. While his equations matched reality when he quantized the energies involved, he didn't have a solid theory as to what form the energy was in when it was behaving so discreetly.

The physicist who *did* come up with such a theory needs very little introduction: Albert Einstein. Building on Planck's work, he postulated that the energy in discrete form actually behaved as particles, different in nature from the electrons, protons, and neutrons in the atom. These particles of energy would come to be called *photons*. Einstein used them to explain the mechanisms behind something called the *"photoelectric effect"*: When some materials are struck by light or other electromagnetic radiation, they release electrons (Muller, 2022). Devices such as solar cells, which generate electricity from sunlight, and the "film" inside a digital camera make use of this effect to function. The kinds of energies that the electrons can gain from the radiation are better explained by a discrete form of energy than a continuous one, which is where photons come in.

The fact that electrons absorb or lose energy as photons has had a huge influence on theories about the structure of the atom. Particularly, this helped develop the model of the atom that is still in use to this day.

Before quantum physics truly made a splash, the structure of the atom was thought to be like our solar system: Like horses turning around the center of a merry-go-round or planets revolving around a sun in any number of possible orbits, electrons would revolve around the central nucleus in a continuous spectrum of possible circular orbits. This was also known as the Rutherford model of the atom. The physicist Ernest Rutherford developed this model after his observations from firing alpha particles (a kind of radioactive, heavy particle that is roughly the size of a helium atom) at a very thin foil of gold and observing how the particles scattered (or not) on impact.

The first blow to this model came from a classical physics observation of charged particles: When they accelerate, alpha particles give off electromagnetic radiation, and any object that moves in a circular path is constantly accelerating. You may have felt the effects of this acceleration yourself when sitting in a merry-go-round or when inside a vehicle that is taking a sharp turn. So, according to classical physics, an electron orbiting around a nucleus would constantly give off radiation and lose energy, eventually spiraling into the nucleus at the center of the orbit.

The quantization of energy and its manifestation as photons helped fix this particular model, as it only allowed electrons to absorb or emit energies of specific amounts—so they couldn't continuously spiral into the nucleus anymore. This led to the next model of the atom, known as the Bohr model, named after the physicist who proposed it, Niels Bohr. In this model, electrons still orbited around the central nucleus but only did so at specific orbits, which depended on the size of the photons that the electrons could absorb or lose. In essence, the energies of the electrons inside the atom are quantized as well.

This went some way toward explaining why electrons didn't just collapse into the nucleus after a while, but it wasn't the most satisfying solution. For one thing, the Bohr model couldn't fully explain energy levels and electron behavior in atoms bigger than a hydrogen atom, which is the smallest possible atom that we know of. There were also some other experimental observations that it couldn't account for, like the Stark and Zeeman effects.

The physicist Louis de Broglie suggested an inversion of what had been done for energy years before: In the same way that waves of energy could behave like particles, matter, such as electrons, could behave like waves. This was the property that would come to be *known as "wave-particle duality."* Building on this concept, physicist Erwin Schrödinger devised what the electron in an atom would be like in waveform, or the "wave function" of the electron. In this model, the electron was not a particle in orbit but a wave confined within the atom, just as the waves on a guitar string are confined inside the string. And, just as the guitar string itself doesn't shift position or move while it vibrates, the electron also didn't move, solving the issue of why it didn't constantly lose energy.

This is how the currently accepted model of the atom, the Schrödinger model, came to be—the one where the nucleus at the center houses the protons and neutrons of the atom while the electrons exist in a wave-like form in the cloud around it. As you can see, the development of this understanding of what is inside an atom also gave rise to or was otherwise influenced by several major concepts of quantum physics itself.

Key Physicists in Quantum Physics

Max Planck (1858–1947)

Max Planck, a German physicist, is often regarded as "The Father of Quantum Theory" due to the groundbreaking contributions he made to the field, particularly at a time when classical physics still held sway. It was in 1900 that he first devised the mathematical concept of quantized energy to explain the experimental observations of black-body radiation. At the time, however, he saw it as a desperate "mathematical trick" to fit his formula to experimental observations without having the corresponding physics theory to fully support it—this would come later from his close friend and fellow physicist, Albert Einstein. He was awarded the Nobel Prize for Physics in 1918 for effectively giving birth to an all-new branch of physics.

Albert Einstein (1879–1955)

For someone who was quite influential in the development of quantum physics, the German-born Albert Einstein was extremely skeptical of its implications. In particular, he did not like the idea of indeterminate physics and famously said, "God does not play dice with the universe" (Baggott, 2011). He didn't start out that way, though; he supported theories such as Planck's quantization of energy and de Broglie's wave-particle duality. Despite his later distaste for the field, he was awarded his only Nobel Prize for Physics in 1921 for his contributions to quantum physics through his concept of the photon and his work on the photoelectric effect.

Niels Bohr (1885–1962)

Niels Bohr, a Danish physicist, was the first to incorporate quantum ideas into the structure of the atom, such as quantized angular momentum (i.e., electrons could only rotate at certain fixed speeds around the nucleus) and discrete electron orbits (electrons could only rotate at certain fixed distances away from the nucleus). Bohr was also a key proponent of the indeterminate nature of quantum mechanics. He had a series of public disputes with Einstein on the subject, which would go on to inform many of the accepted views that followed. When Einstein gave his famous dice quote, Bohr's response was, "Einstein, stop telling God what to do" (Baggott, 2011). Bohr received the Nobel Prize for Physics in 1922 for his work on the atomic model.

Louis de Broglie (1892–1987)

In 1924, French physicist Louis de Broglie proposed wave-particle duality after observing how X-rays interacted with electrons through the photoelectric effect in the lab of his older brother and fellow physicist, Maurice. It was an extension of the idea of photons, i.e., waves that behaved like particles under certain conditions. Born to an aristocratic family, de Broglie spent his youth studying on his own and even fought in World War I effectively on his own as he operated radio equipment inside the Eiffel Tower. He was awarded the Nobel Prize for Physics in 1929 after his theory of wave-like matter was proven experimentally.

Werner Heisenberg (1901–1976)

Together with Bohr, the German physicist Werner Heisenberg was a key supporter of the indeterminism of quantum physics. He made many contributions to the field of quantum mechanics, the underlying fundamentals of quantum physics as a whole—the most famous of these contributions being his uncertainty principle, which, in brief, states that certain pairs of properties, like the position and speed of a particle, can only be measured simultaneously up to a limited amount of accuracy. It became a foundational principle of the field, and he was awarded the Nobel Prize for Physics in 1932 for "the creation of quantum mechanics."

Erwin Schrödinger (1887–1961)

Independent of Heisenberg, the Austrian-born Erwin Schrödinger worked on wave mechanics, which was another side of the coin: quantum theory. While he is most famous for his thought experiment involving a cat in a box, his primary contribution to quantum physics was the Schrödinger equation, a mathematical framework that describes how quantum systems operate. He was also one of the first scientists to contemplate the concept of quantum entanglement and even coined the term himself. He was awarded the Nobel Prize for Physics in 1933 for his efforts.

Paul Dirac (1902–1984)

Paul Dirac was the other winner of the Nobel Prize for Physics in 1933, also for his contributions to quantum mechanics. In particular, Dirac named and established the field of quantum electrodynamics, which describes how electromagnetic radiation, like light, interacts with matter particles like electrons and protons. This field is also notably tied to Einstein's theory of special relativity, which, together with quantum mechanics, moved science a little closer to the elusive grand unified theory that explains everything.

TWO

Wave-Particle Duality

Double-Slit Deliberations

I f you've been to a concert at a badly designed music hall, you may have found yourself in a dreaded "dead spot": a place where the music from the speakers doesn't sound as loud or clear as it should. Or, if you have a microwave oven in which the plate doesn't rotate, you might notice your food being more unevenly heated than it should be.

Both of these are examples of a property that waves have called interference. All waves travel by causing something to vibrate back and forth: sound waves travel by vibrating the air particles or other material that they travel through, and microwaves travel through the vibration of electric and magnetic fields within the wave itself. This is why a microwave and other waves of the same family, such as light and X-rays, are called electromagnetic waves or radiation; they all travel by using the same mechanism.

So, when a wave travels, the object that is vibrating in the process can do so in one of two opposing directions. Let's call these two directions "up" and "down" for the moment (the actual directions depend on where the wave is traveling, as well as what type of wave it is). When two waves of the same kind meet each other, they are able to *superimpose*—simply put, the direction of their vibrations combine at the point where they meet. If both vibrations are in the same direction (i.e., two ups or two downs), then they add together, and the resulting vibration is doubled in size. This is called *constructive interference*. If the vibrations are in opposite directions (i.e., an up *and* a down), they try to cancel each other out, and the resulting vibration is reduced in size. This is called *destructive interference*.

Those dead spots at the music hall are caused by the sound from the speakers interfering destructively in those areas, which is why the resulting sound is muted. This effect may not be desired at a concert, but it has its uses. For example, noise-canceling headphones make use of it by reducing the volume of ambient sound outside the headphones. As for the microwave oven, the waves emitted by a microwave oven form a pattern of points inside the appliance with both constructively and destructively interfering waves; the food is heated more at the former and less at the latter. This is why microwave ovens usually come with a rotating plate inside, so that the food gets moved through all the various interference points and all the parts of the food are equally heated over time.

You would not expect this kind of behavior from particles, though. If two tennis balls are thrown at each other, they don't combine into a bigger tennis ball where they meet and then carry on along their paths from there! And yet, something not unlike this is what occurs at the quantum level of reality and is the core of the idea of wave-particle duality. And one of the key experiments that confirmed this was the double-slit experiment.

So, what is a double-slit experiment in the first place? To understand how it works, think about how light comes out of a flashlight. If you shine the flashlight on a wall, you will see a bright bit at the center, which fades out as you move further away from the area the flashlight is pointed at. Now, imagine that you place a sheet of paper between the flashlight and the wall, and on the sheet of paper is a single thin slit, about a millimeter in width. Light would only go through the slit, and you would see a slit-shaped patch of light on the wall, even though there would still be some faint light fading out at the edges of the patch.

So, what happens if there are two slits instead of just one between the flashlight and the wall, placed close to each other and parallel? What kind of pattern forms on the wall? Well, that depends on whether light is a wave or a particle. You know from the earlier examples in this chapter that two sources of the same type of waves will form an interference pattern when those waves interact with each other. This would show up as alternating bright and dark patches on the wall, which is an *interference pattern*—that is if the slits are as thin as the wavelength of the light itself. The reason for this is another property of waves called diffraction, which is a whole other topic in itself. If the two sources are particles, though, then the effect would be something like two cans of spray paint painting next to each other on the wall: The light would slightly overlap in the middle, and there would be two bright spots where the light from each slit is directly hitting the wall, and then the spray fades out over distance.

Usually, light that comes out of a source like a flashlight is composed of many different wavelengths of light, is emitted at different times, and is scattered out in many directions—this is why light that comes out of a typical flashlight or the headlights of a car will form a roughly conical shape. This type of light doesn't show a very clear interference pattern during double-slit experiments, which is why laser lights are often used instead. Laser light is specially treated so that it has only one wavelength and is emitted coherently, meaning it happens at the same time and in a specific direction. If you have ever used a laser pointer, this is why it forms a very bright dot on the wall instead of a diffused cone. The resulting interference pattern from the laser light is also much clearer and easier to analyze during a double-slit experiment.

The question of whether light is a wave or a particle was what British polymath Thomas Young tried to answer in 1801 when he carried out the first double-slit experiment. At the time, there were two trains of thought concerning light: The famous Sir Isaac Newton believed that light was made up of tiny particles, which he called "corpuscles," while Dutch scientist Christiaan Huygens believed that light was made up of waves. During that experiment, the light from the slits formed an interference pattern, and for the time being, Huygens was proven to be right. However, as you read in the previous chapter, light can also behave like particles under certain conditions—Newton wasn't completely wrong!

The Rise of the Matter Wave

In the 1920s, at a time when quantum physics was still finding its stride, two physicists, Clinton Davisson, and Lester Germer, were using electrons to carefully study the surface and structure of materials like nickel and other crystalline solids (solids that have a very regular internal structure, like ice as opposed to clay).

Inspired by how Ernest Rutherford investigated atoms by firing particles at a sheet of thin gold foil, Davisson and Germer hoped to analyze the internal structure of crystalline solids by firing a beam of electrons at them and then observing how the electrons reflected and scattered off their surface. They began in 1924 by firing electrons directly at a nickel crystal (or perpendicular to its surface) and using a movable detector to find where the electrons were scattered.

The experiment was supposed to be conducted in a vacuum so that the electrons wouldn't be disturbed by air particles, as it would have affected their paths and given a distorted view of the nickel crystal's structure. However, during the experiment, air accidentally entered the chamber and caused the nickel to form a layer of nickel oxide on its surface. Since the structure of nickel oxide is different from that of nickel, this needed to be removed, which they did by heating the crystal to a high temperature.

Unknown to them, this also altered the structure of the nickel crystal itself, and the resulting solid had atoms that were regularly spaced at what turned out to be a very important distance from each other. When they fired the electrons at the solid after the oxide layer had been removed, they were stunned to find that the electrons were now scattering in alternating sections of high and low intensity, according to the detector. In essence, the electrons were displaying an interference pattern.

You can observe this kind of reflected interference pattern by looking at the uncovered surface of a CD, DVD, or an oil puddle; the tiny grooves on the discs and the particles inside the oil are spaced apart at roughly the same wavelength as light itself. So, when light reflects off these surfaces, the effect is like light being passed through the two slits in a double-slit experiment. The reason you see so many different colors is that each of the colors in natural light has a different wavelength, so they interfere constructively and destructively at different points. When a particular color, such as red or blue, interferes constructively at a point, that is the shade you see at that point.

Davisson and Germer simply marked their electron observation off as an anomaly and restarted the experiment with a fresh nickel crystal. In a remarkable coincidence, another physicist by the name of Louis de Broglie was independently proposing something radical in his Ph.D. thesis around the same time: That a moving particle of matter could behave like a wave, which he appropriately called a *matter wave*, with a corresponding wavelength and other properties normally associated with waves.

When Davisson and Germer had heated the nickel crystal, the atoms were spaced apart at a distance that corresponded to the matter wavelength of the electrons at the speed at which they were fired at the crystal. Due to this, they were scattered much like how light is scattered off the surface of a DVD, which is why they formed an interference pattern. The two physicists had, basically, proven de Broglie's theoretical musings to be right. After de Broglie's concept gained more attention, they repeated their experiment a few more times to confirm the interference pattern of the electrons. By 1927, they confirmed that electrons had a corresponding matter wave that they behaved like under certain conditions. The duo would go on to receive Nobel Prizes for their efforts in 1937.

The weirdness of the wave-particle duality on display didn't stop there, though. And, to understand this further requires a return to the double-slit experiment. What follows is a simplification of the results observed when electrons were subjected to the above experiment.

Imagine that the flashlight or laser light from earlier is now replaced by an electron gun, and it's now electrons that are being fired through a single slit (for simplicity, let's also say that the wall past the slit has been treated with a special chemical that shows where electrons land on it—electrons are otherwise invisible, after all). Much like how paint from a spray paint can act when sprayed through a stencil, the electrons go through the slit and land on the wall the way particles would: an intense spot in the center that fades in intensity as you move away from it. So, with a single slit, the electrons behave like particles.

Now, replace the single slit with two slits. What happens then? Lo and behold, the electrons form an interference pattern on the wall. They are now behaving like waves. The same pattern is eventually observed even if you fire the electrons from the gun one at a time so that they go through the slits in a single-file line, as it were. This, if you think about it, is a hint of quantum weirdness; if the electrons went through the slits one at a time, how could they interfere with each other?

To solve this issue, a detector can be placed on the sheet with the slits to see which slit the electrons move through as they travel past the sheet. This is where more quantum strangeness occurs: The detector shows that roughly half the electrons travel through one slit, and the other half travel through the second slit, which makes sense. But, when the detector is active, the electrons no longer show an interference pattern on the wall but two particle-like splotches instead—which doesn't make nearly as much sense. Somehow, actively monitoring the electrons causes them to behave like particles again.

A reasonable assumption at this stage is that maybe the presence of the detector is what causes the change. To test for this, the detector can be left in place but switched off, and the experiment is repeated. What happens then? As it turns out, the electrons form an interference pattern on the wall, behaving like matter waves once again.

These double-slit experiments on electrons demonstrated one of the core peculiarities of quantum physics that still stumps scientists to this day and age: *the act of observation of particles and waves at the quantum level changes their nature.* They also confirmed one other very important foundational principle of quantum physics: Just as how light and other electromagnetic waves could behave like particles when observed under certain conditions, moving particles could behave like traveling waves analogously.

Wave-Particle Duality in Practice

Along with completely rocking the foundations of our beliefs concerning what waves and particles are, wave-particle duality has also lent itself to many practical applications in various technologies, innovations, and natural phenomena. One such example is electron microscopy.

If you've used a microscope before, you have some idea of how it functions: It allows you to see things that cannot normally be seen with the naked eye, such as the surface of a hair, certain types of tissues in the body, or small animals like bacteria and amoeba. The word "microscopic" is based on this device; if something can only be seen clearly under a microscope, then it's considered to be microscopic. However, standard microscopes are limited because they are optical by nature—that is, they make use of light reflecting off the object that is being magnified so that we can see it.

To understand why this is an issue, consider how we view things in the first place. We see a book because light bounces off the book before entering our eyes. We see specks of dust dancing in a beam of sunlight because the sunlight is bouncing off those tiny little dust motes. We *don't* see the air molecules (the particles that air is made up of) that are smacking those dust motes around, though, and this is because the wavelength of light is too large for it to bounce off the air molecules. What happens instead is something like water rushing around a small rock in a river; the light simply moves around the air rather than reflecting off it.

At certain speeds, the wavelength of the corresponding matter-wave of an electron is short enough to be able to reflect off even the smallest of molecules. Then, with the aid of special detectors, the scattered electrons from the target in the microscope can be converted into a viewable image. This allows magnifications of an even higher resolution and detail than optical microscopes can provide.

Electron microscopes have enabled scientists to examine objects at the atomic and molecular levels, which has provided some extremely useful insights into the structure of materials, samples of biological systems like the roots of plants or the surfaces of insects, and complex chemical compounds. As you can imagine, this technology has been crucial in the development of various scientific disciplines, ranging from materials science to biology.

Another type of microscope takes things even further by making use of an additional quantum effect called tunneling (which will be covered later in this book). Scanning tunneling microscopes, or STMs, use electrons to observe and visualize objects and surfaces at an even smaller scale. They achieve this because when electrons behave like waves, they have the ability to perform the phenomena of tunneling, something that particles cannot do while behaving like particles.

Outside of the more practical applications, though, the concept of wave-particle duality was one of the first causes of the shift in physics from a foundation of determinism to one of indeterminism. It also led to some spinoff interpretations that continue to be debated to this day, such as the Copenhagen interpretation and the Many-Worlds interpretation. The former places a focus on the act of observation and states that the reality we live in is effectively created when we observe it, and it is in a sort of limbo of superimposed states until then (much like Schrödinger's Cat). The latter interpretation suggests that at the moment of observing a system, it splits into parallel co-existing universes in which all the possible outcomes of the observation occur. So, in the case of Schrödinger's Cat, observing the animal results in at least two groups of separate universes: in one group, the cat is alive, and in the other, it is dead.

So, while it heralded a rather uncomfortable shift in attitudes toward the subject for many physicists, wave-particle duality also enabled us to see so much more of the inner workings of the world than classical physics on its own could provide.

THREE

Heisenberg's Uncertainty Principle

The Quantum Limits of Accuracy

Measurements are a cornerstone of science, and it's not an exaggeration to say they're fundamental to our daily lives and routines, too. We constantly look at the time to determine what we must do next. When we decide what to wear, we choose clothes that have been measured and adjusted in length and shape to make us look a certain way. When we buy groceries at the supermarket, we're constantly looking at weights and prices so we know just how much we need to get.

Take any one of these measurements and imagine the chaos that might occur if they were to become more unreliable all of a sudden. How would your morning routine be affected if the time on your clock was anywhere from 10 minutes behind to 10 minutes ahead at any moment that you looked at it? How would grocery shopping feel if the items you bought had different weights and costs, depending on how and when you observed them?

The actual implications of the uncertainty principle are something like the above, although not quite as dramatic. Still, it might give you some idea of just how much physicist Werner Heisenberg rocked the foundations of physics when he introduced his concept to the field at large. So, how exactly does the uncertainty principle affect measurements? To understand this, let's consider the following thought experiment:

> Imagine a game board on which there are some marbles. The game board is a representation of the space in which particles exist (something like the electron cloud, where the electrons are in an atom), and the marbles are the particles themselves. The board is very smooth, so the marbles are free to move around the board in any direction.

If you've played a game of marbles before, you know that there are two very important properties of the marbles as far as the game is concerned: Where the marbles are and where they're going. The same applies to our marbles on the board here—the properties of the marbles we want to measure are position and speed.

How are positions measured at the scale of reality you're more used to? Or, simply put, how do you normally find where things are? The obvious answer is that you look for the thing in question. And, if you need it, you can sometimes use a light source like a flashlight for this. Physically speaking, what you're doing is looking for photons of light reflecting off the target of your search. When those photons enter your eyes, you subconsciously trace them back to where they came from and then find the item you were looking for. More precisely, you find its position.

The same applies to the quantum level of reality: The position of particles is found by emitting photons in their vicinity and then detecting the photons that reflect off the particle we're looking for. So, to find the position of a marble on the board, you would send photons toward the board and look for the ones that reflect off the marble.

However, remember that photons are packets of energy. At the quantum scale, these packets of energy are roughly the same wavelength as the particles they are used to observe. What happens when you strike a marble with a packet of energy? It gains some amount of energy and *moves at a different speed from before.*

From the workings of electron microscopes, you might have deduced that the smaller the wavelength of the particle used to observe something, the more detail and accuracy you can obtain in the final image. The same applies here: photons of a smaller wavelength lead to more accurate measurements of position. However, the wavelength is inversely related to energy, meaning photons of a smaller wavelength transmit a greater range of possible energies to the particle they're reflecting off.

This is the core of what Heisenberg proposed with the uncertainty principle: *The more accurately you attempt to measure the position of a particle, the less accurate your information about its speed becomes, and vice versa.*

To see the inverse of this effect, consider the marbles on the game board again. To measure their speed, you can place a tiny revolving door that the marbles can collide with on the board. By looking at how quickly the door spins from the collision, you can measure how quickly the marble was moving when it struck the door. You might have already spotted the issue here, though: the collision causes the marble's position to shift. And, just like before, the more robust and accurate the revolving door is at detecting the marble's speed, the more it will cause the marble's position to shift.

A similar effect can be observed in waves, with two different properties: the wavelength of the wave and its position in time. To measure the wavelength of a wave accurately, it needs to be observed over a period of time, which means that the measurement of that time becomes less accurate. To put it another way, we are less certain about which moment in time the wavelength of the wave was the one we measured. Conversely, if a wave is obtained for only a brief instant of time, then what can be determined about the wave's wavelength becomes less accurate.

The uncertainty principle, therefore, places a cap on how accurately we can measure these pairs of properties at a quantum level. This is in part due to a peculiarity that was also revealed when discussing wave-particle duality in the previous chapter: *the observation of a thing at the quantum level causes the nature of the thing itself to change.* Our methods of observing particles at a quantum level, which needs to be done in order to make measurements, cause their properties to change. The more accurately you attempt to measure one of these properties, the greater the change you cause to the associated properties, which means your information about them becomes less accurate as a result.

This also has another implication when it comes to these properties: that they are *inherently uncertain.* This is to say that, even before making an observation or measurement, the properties of a quantum system are not set in stone. If, for example, there is a range of possible positions that a particle can have in a quantum system, then until it is observed, it has a non-zero chance of being in any one of those different positions at any given moment—and the same applies to all of its other properties that are governed by quantum laws.

Uncertain Cryptography

In communication, especially in our modern day and age, it is vital to make sure that messages and information can only be read by the people they're intended for. You wouldn't want the wrong people to be able to hack into your emails or text messages or be able to extract data about your bank accounts and online profiles. Cryptography, or encryption, is a process in which information is protected by obscuring it while it is sent from one point to another, giving only the intended recipient the means of revealing the actual information (also known as "decrypting" the information).

Cryptography will be covered in more detail in Chapter 8. However, the technology behind quantum cryptography makes use of Heisenberg's uncertainty principle to function, and that is what will be examined here.

Quantum cryptography, although still in its early stages of development, makes use of the fact that the properties of particles are uncertain by nature. Not only is it not possible to determine all the properties of a particle at any given moment with a high level of accuracy, but the nature of the particle itself can change based on the observation (as implied by wave-particle duality). Or, put another way, properties at a quantum level cannot be measured or observed without causing changes. This is what gives rise to the uncertainty principle: the means used to increase the accuracy of one kind of measured property cause changes and uncertainty in other properties as a result.

A type of cryptography that uses the above in its mechanism is one known as *quantum key distribution*, or QKD. More specifically, it makes use of the fact that any attempts to observe a quantum system cause changes to the system in the process. So, how does that factor into cryptography?

Consider the following scenario:

> Two people, Alice and Bob, want to establish a secure connection between them for exchanging messages and other information. They can do this by setting up a communication channel between them that makes use of optical fibers—these are very thin glass or plastic fibers that can transmit photons within them from one point to another.

Photons are particles that follow the principle of quantum physics. This means that, as the uncertainty principle shows, if any property of the photons is observed or measured, the act of doing so causes changes and uncertainty to their other properties in the process.

Going back to the scenario with Alice and Bob, a third party named Eve tries to hack into their communication channel. To do this, Eve would need to observe or measure the photons being used in the connection—and this is where quantum physics rears its head. By making those observations or measurements, Eve would inevitably cause some kind of change to the photons themselves. In a QKD system, these changes can be detected by Alice and Bob, immediately letting them know that their communication channel has been intruded on.

Of course, that's not all there is to QKD; it also makes use of the property of quantum entanglement in its functioning, and there are many issues left to overcome in order to practically implement it. These other aspects will be discussed in detail later.

Philosophical Ramifications

The uncertainty principle, along with its implications, was not only a shock to the field of physics but also to the realm of philosophy, particularly to the determinism worldview that most scientists had at the time. After all, in the classical physics framework that had dominated our understanding of reality for centuries, if the present conditions of a system were accurately known (such as the mass of an object, its temperature, the presence and location of other objects, and so on), then it was possible to determine future conditions with just as high a level of precision. For example, if a box is dropped from a height, and everything about the box is known (such as what it is made of, where it is, and what else is below it), then what happens to the box next can be predicted very easily and accurately.

In a way, physicists saw the world as a sort of clockwork mechanism the size of the universe itself, where everything that happened did so according to various deterministic laws of physics. The uncertainty principle not only threw a wrench into that clockwork but threw one that couldn't even be detected with a satisfactory level of accuracy.

Take, for instance, the fact that, at the quantum level, the act of observation influences the thing being observed. This concept clashed with the idea that anything that exists does so independently of the person observing it. In the reality of classical physics, a red ball remains a red ball whether you interact with it or not, and a tree falling in a forest makes a sound whether anyone hears it or not since the tree follows laws of physics that predict it will do so. The quantum perspective, therefore, raises questions about just how objective our reality is and how much of it is influenced by our conscious observation of it.

As is to be expected, whenever an idea that threatens our understanding of reality is introduced to the world, many debates also arise regarding the true nature of particles. One of the more famous series of these debates took place in the 1920s and '30s between two notable physicists, Albert Einstein and Niels Bohr. Their opposing views on quantum mechanics gave rise to some profound questions about the core nature of the universe.

Bohr was in favor of a probabilistic view of quantum mechanics, which is why he proposed the Copenhagen interpretation of reality, which states that particles at the quantum level exist in multiple possible states at the same time. These particles "collapse" into a single, definite state only when the particle is observed. To this end, Bohr also placed a lot of importance on the idea of *complementarity*, which is another way of expressing the idea of wave-particle duality that particles could exhibit both wave-like and particle-like properties, depending on the experimental context.

Einstein, on the other hand, was deeply uncomfortable with this probabilistic and uncertain view of quantum mechanics. He advocated for a more deterministic perspective instead and, to support this, proposed the existence of hidden variable factors that were behind the apparently random results of quantum events or the fact that they could be made more uncertain through measurements.

The debates between Bohr and Einstein, which went on for several years and were ultimately never definitively resolved, were not only scientific but also philosophical—and, true to both of their intellectual natures, these debates were very civil and respectfully conducted. Einstein challenged the completeness and consistency of quantum theory by raising concerns about the lack of a clear, objective reality at the quantum level. In response, Bohr defended the notion that quantum mechanics even needed to provide such a comprehensive, accurate picture of the microscopic world, even if it meant departing from classical ideas of physics.

Aside from these debates, the uncertainty principle has also sparked questions about the limits of human knowledge. If there are limits to the precision with which we can know certain properties of particles at a fundamental level, it raises doubts about the extent to which we can truly understand the universe. This goes beyond the realm of physics and touches on considerations about the theory of knowledge itself, as well as the nature and scope of how much humans can know and learn.

There are also implications for the concepts of destiny and free will. After all, if the behavior of particles can be so fundamentally uncertain and unpredictable, then what does that say about the behavior of people? The uncertainty principle challenged the view that causes and effects are predetermined and predictable while supporting a more nuanced take on the ideas of agency and choice.

You can see, by now, how the ideas of wave-particle duality (or *complementarity,* as Bohr called it), together with Heisenberg's theory, had such a profound impact on determinism, objectivity, and the nature of reality itself. Even to this day, these concepts continue to fuel discussions about the universe, our role in it as observers, and the limits of how much we can know and understand about our existence.

FOUR

Quantum Superposition and States

The Cat in the Box

When most people hear about Schrödinger's Cat, they imagine a cat inside a box that is both alive and dead at the same time and usually leave it at that rather weird bit of imagery. However, not only is Erwin Schrödinger's thought experiment a little more detailed and complex than the above description, but its implications for how particles behave at the quantum level extend far beyond being merely peculiar.

Ironically enough, although Schrödinger's Cat became an iconic pop-culture representation of the unpredictable nature of quantum physics, Schrödinger himself was not a supporter of the more indeterministic views that dominated the field in later years. His thought experiment was meant to ridicule the supposed uncertainty of quantum reality, not become its symbol.

The first step to understanding the ramifications of the experiment is to look at its setup. There is the box: when sealed, its contents cannot, in any way, be observed from the outside of the box. In other words, the cat and the other contents within cannot be imaged by an X-ray scanner, for instance. The only way to observe what is inside the box is to open it. Inside the box are placed the following: the cat, a vial of fast-acting poison called hydrocyanic acid, a small hammer connected to a Geiger counter, and a small piece of radioactive material placed near the Geiger counter.

Radioactive materials are radioactive because they are in an inherently unstable state and go through a process known as *radioactive decay* to become more stable. Recall the structure of an atom: It has protons and neutrons in the nucleus at its center and electrons in a cloud around the nucleus. For every number of protons that the nucleus can have, there is a corresponding number of neutrons that result in the most stable configuration of that nucleus. As an example, a carbon atom has six protons, and in its most stable form, it also has six neutrons. However, the element carbon isn't found naturally in this form only; sometimes, carbon atoms have eight neutrons instead of six. The resulting less stable atom, carbon-14, is what's called an *isotope* of carbon. Radioactive materials are always made up of isotopes like this one.

If you place a typical broomstick upright in the middle of a room and let go, it will fall until it's lying on the ground. Physically unstable objects naturally try to shift to a position of greater stability, and the same is true for isotopes, which is why they undergo radioactive decay. While the detailed mechanisms behind decay are very complex, for now, it is enough to know that during decay, the unstable isotope gives out a small particle of energy from its nucleus to become more stable. This process is random in nature, like the outcome of a coin toss; the moment when an isotope decays cannot be predicted.

Geiger counters are devices that can detect the small particles and energies emitted by radioactive material. The one in the thought experiment is configured so that when it detects emissions coming out of the radioactive material, it triggers the hammer to smash open the poison vial, thereby killing the cat instantly.

After the contents are placed inside the box—with some protection in place so that the cat cannot accidentally smack the rest of the equipment inside, as cats tend to do—the box is sealed. The thought experiment takes effect after about an hour: The radioactive material *will* decay and be detected by the Geiger counter at some point, resulting in the cat being poisoned to death; however, the exact moment cannot be predicted. This means that, at any time, the state of the cat (whether it is alive or dead) is uncertain.

Why the complicated setup for this thought experiment? Since this is quantum physics we are dealing with here, particles of a quantum nature need to be featured—this is the purpose served by the radioactive material, whose isotopes are quantum in nature. The rest of it is a setup meant to strongly link the state of the cat to the state of the radioactive isotopes: When they decay, it dies, and until then, it is alive. The effect is that a big, macroscopic object like a cat (by particle standards) suddenly takes on very quantum properties, which is why it can now be treated as a quantum object itself.

When two waves of the same type meet at a point, they can combine either constructively (adding on to each other) or destructively (opposing each other), depending on their state at the meeting point. But the wave itself is not irreversibly altered in the process—beyond the meeting point, the waves continue as they did before. This can also be called *superposition*: At the meeting point, the two waves exist in the same place at the same moment, and what is observed is the result of them combining since they share the same space and time. This doesn't apply to only two waves either: When you look at a light being reflected off a DVD surface, the result is the superposition of several different wavelengths of light.

Quantum superposition takes the idea of several states combining and applies it to the uncertain nature of particles. It combines ideas from wave-particle duality and the uncertainty principle to state the following: *a quantum system exists as the result of several possible states of that system superimposed on one another.* Going back to the cat in the box, the condition of the cat can be either one of two possibilities: alive or dead (if you want to be creative, you could add a few, such as the cat being on the verge of dying, experiencing the first signs of poisoning, and so on. However, these are all transitional states that do not last very long compared to the cat being either alive or dead). From a quantum superposition perspective, the cat exists as these two states superimposed on each other—this is why the cat is both alive *and* dead.

Recall that the act of observing a quantum system changes the nature of the system itself. The same applies to the cat: the only way to observe its state is to open the box and look inside. According to most interpretations of the experiment, doing so causes what is usually called a *collapsing* or resolution of the states; basically, the system takes on the traits of any one of its possible states. In the case of the cat, when the box is opened, and the cat is observed, it will either collapse into an "alive" state or a "dead" state.

The more likely any state is to exist, the more likely it will be the one chosen. In the example of the cat, it is more likely to be observed as alive earlier on in the experiment and more likely to be observed as dead after several hours. Another way of saying this is that the quantum condition of the cat is a combination of various probabilities or the chances of it existing in a certain state. The physics term used for such a state is a *probability function*. One of the more important probability functions in quantum physics is one we've already encountered before: the wave function of an electron inside an atom.

Devised by Schrödinger himself, the wave function of an electron is a mathematical expression of how the electron in an atom behaves when in waveform. And, like waves that you might be more familiar with (such as the ripples on the surface of a pond), the waveform of the electron also has peaks and valleys at various points along the space it occupies. However, they're not arranged in quite the same way as they might be in your typical wave.

So, what does this have to do with quantum superposition and probability? You might have guessed the first one already: the wave function of an electron is a result of all its possible states as a wave superimposed on one another. As for the second one, think back to the peaks and valleys. In an electron's wave function, the peaks (or points of high *amplitude*, a measure of a wave's size or magnitude) represent the points where its position is more likely to be when it behaves as a particle, and the valleys (or points of low amplitude) are where it is less likely to be located. In other words, if the electron is observed in a way that causes it to collapse into particle form, the amplitude of the wave function shows the chances of finding the electron in this form at any point within the space it occupies.

From Quantum Computers to Quantum Plants

Computing is a technology that is built on a foundation of bits. Bits are the spaces in which information is created, stored, processed, or exchanged, and information is everything in the world of computers. The commonly used acronym for it, IT, *does* stand for "information technology." Due to their basis in electronic circuitry, bits can classically take on two values: 0 (or "off") and 1 (or "on").

Quantum computing is still in its earliest stages, but it promises to revolutionize computing by altering it at this very fundamental level. It does so by merging the principles of superposition and quantum entanglement with the structure of the bit, resulting in the *quantum bit*, or "qubit" for short.

Whereas a classical bit can only be either 0 or 1, a qubit can be a superposition of both or multiple, in which there are different probability functions for the 0 and 1 states. This means that a qubit is able to store and make use of vastly more information than a standard bit can. To see this in action, consider what two bits can store, which would be one of the following four combinations: "00," "01," "10," and "11." Or, put another way, you would need four sets of these bit pairs—eight bits in total—to store all four possibilities. But, with qubits, thanks to superposition, two qubits are enough to store all four combinations at the same time. If you add one more qubit, what the three of them can store requires 24 classical bits—and this difference grows exponentially larger as you increase the number of qubits.

Since qubits can store more information, they can also perform calculations and solve problems with that information much faster as well. As a quick example, take the mathematical problem of finding the prime factors of a big number (a prime factor of a number is a prime number that can divide the original number without leaving a remainder. For instance, the prime factors of 50 are five and two). A computer using standard bits would take millions of years to find the prime factors of a number that is 600 digits long. However, a quantum computer, using qubits, can solve the same problem in minutes.

Computers are not the only field in which quantum superposition can support a boost in efficiency. As it turns out, nature had been making use of this phenomenon long before we discovered it. Plants and some types of bacteria make use of a process called *photosynthesis* to take in energy from sunlight and convert it into stored chemical energy. This chemical energy is similar to the energy we get from our own food and drink in that it allows these plants and bacteria to perform the functions they need to perform in order to live.

The process of photosynthesis makes use of special molecules that can absorb light, called *chromophores*. These chromophores can exist in a superposition of different states during photosynthesis, which allows them to absorb the sun's energy through multiple different pathways at the same time. Similar to the qubits and quantum gates above, this allows the process of photosynthesis to be performed much faster and more effectively.

Future Possibilities

Quantum computing, if it can be advanced to a stage at which it has overcome its current practical issues and become a bigger presence in our daily lives, is just the tip of the iceberg for applications of quantum superposition in technology.

For example, take the issue of data encryption that we touched on in the previous chapter. Encryption could make use of quantum superposition to transmit the data in a superposition of different states, making it much harder for someone outside of the transmission to be able to extract the information they want. Or, the algorithms that are used to obscure the data could make use of quantum computing, which is exponentially harder to break through.

Encryptions aren't the only thing about those data transfers that could be improved—their speed and bandwidth, or the amount of information that can be sent through a communication channel at any given moment, could also be increased dramatically. Imagine having blazingly fast, highly secure ways to communicate throughout the globe, all facilitated by quantum superposition and other quantum properties like entanglement. Such communication networks could see great advancements in all manner of fields, from finance to healthcare, by enabling people to work together and make important decisions more effectively and on a much larger scale than before.

Another potential use of quantum superposition is in the development of sensors (devices used to detect things in the surroundings, such as heat, light, movement, and more) with greater sensitivity and accuracy. These quantum sensors would be able to detect the smallest of changes in things such as gravitational fields, magnetic fields, and even the properties of dark matter—the matter in outer space that we don't know anything about other than the fact that it must exist. Sensors like these could revolutionize fields like geology, astronomy, and even medical imaging—which uses magnetic fields and electromagnetic radiation extensively—leading to more discoveries about the yet-unsolved mysteries of the universe and the human body.

When it comes to research, scientists are trying to use superposition to make use of quantum phenomena at a macroscopic scale—not very far removed from giving cat-in-a-box quantum properties! It pushes the boundaries of what's possible with current technology in many different ways, from the development of new kinds of materials to the improvement of existing devices, enhancing what they're capable of. One particularly exciting area of research is the quest for room-temperature superconductors.

A conductor, in this context, is a material that can conduct electricity—copper wires, steel coils, and the like. Most conductors also have a property called *resistance*, which works against their ability to conduct electricity. Resistance is what causes wiring and electronics to heat up when they've been left on for some time. A superconductor is a material that has zero or negligible resistance.

With our current level of technology, superconductors can only effectively function at very low temperatures, hundreds of degrees below the freezing point of water. The dream, though, and something that quantum superposition could be a key to, is the creation of room-temperature superconductors. Such a breakthrough could herald a new era, not just for electrical power transmission but also for transportation, energy storage, and exponentially more efficient electrical and electronic devices, which would also be far more sustainable and reliable.

FIVE

Quantum Tunneling

The Mines of Moria

In the classic fantasy movie *The Lord of the Rings: The Fellowship of the Ring* (or the book it is based on, depending on your order of exposure to the story), there is a pivotal scene in which the fellowship in question attempts to cross the Misty Mountains. Due to the treacherous terrain and other difficulties, the group resorted to getting to the other side by taking a path that goes underneath the mountain range through an underground complex called the Mines of Moria. In this way, they are able to overcome the obstacle, although, in the story, there are some struggles and sacrifices made along the way.

To understand what any of this has to do with quantum physics, think back to Heisenberg's uncertainty principle for a moment and how it introduces the idea of certain pairs of quantum properties having greater or lesser uncertainty, depending on how they are observed. One such pair of properties for a particle is its energy and the time period in which it is observed: The more accurately the energy of the particle is measured, the less certain the observer can be about the moment in which the particle had this amount of energy. And, the smaller (or more accurate) the period of time in which the particle is observed, the greater the uncertainty about its energy.

Now, picture the following:

> A soccer ball is slowly rolling toward a hill. The hill is about 10 feet high, fairly steep, and very broad. The ball can't simply roll along the side of the hill or smash it out of the way. Can the ball slowly roll over the hill? According to classical physics, it cannot.

While it's rolling on the ground, the ball does not have enough *kinetic energy* (the energy a moving object has) that can be converted into *gravitational potential energy* (energy associated with height and lifted objects, sometimes called *potential energy* for short) to be able to roll up the barrier that is the hill. It can be given this energy by a strong kick, though, which would boost its speed and kinetic energy enough to roll over the hill and onto the other side. Otherwise, it would simply come to a stop somewhere on the slope of the hill and roll back down again.

This is an illustration of what happens to an energized particle when it comes across what is called an *energy barrier* or *potential barrier*. These are barriers in the form of electric fields, magnetic fields, or other forms of physical obstructions that require a minimum amount of energy to be overcome. If the particle does not have this required amount of energy, then it should not be able to get past the barrier in its path. Not according to classical physics, at any rate. But, at the quantum level, this changes.

If the soccer ball was operating at the quantum scale, then when it slowly rolls toward the hill while nobody is looking at the ball, there is a non-zero chance that it will roll past the hill with the hill remaining intact. The effect is as though, like the Fellowship of the Ring using the Mines of Moria to get past the Misty Mountains, the ball has somehow tunneled underneath the insurmountable hill and gotten past it in this way since it did not have enough potential energy to go over it. It, of course, has not done such a thing, but this is where the effect gets its name from: *quantum tunneling*.

If the ball isn't building any tunnels, holes, or detours of any kind, then what does it actually do? To answer this question requires revisiting some of the quantum concepts that have been covered thus far.

Quantum particles have a corresponding waveform, or matter wave, that they can behave as under certain conditions. In this form, the wave can also be referred to as a wave function or probability function, where the amplitude of the wave function gives the likelihood of finding the corresponding particle at that point. The state of a quantum particle is a superposition of all its possible states at a given moment until it is observed and then collapses into one of those states. One way of visualizing all these different states is to imagine that the quantum soccer ball is, actually, several hundred different balls rolling toward the hill at the same time and place.

So, when the ball approaches the steep hill, it is not a single ball with a single amount of insufficient energy; it is, instead, a superposition of different states of the ball, each with different amounts of energy that can all be represented in the form of a probability function. And, given that there is a range of possible energies that the ball can have, there is a non-zero amount of these possible balls with enough energy to roll over the hill and to the other side.

So, what happens? As long as the ball is not observed so that it collapses into a single state, then it is possible states with higher energy will roll over the hill and continue moving on the other side while the remaining states will come to a stop somewhere on the slope of the hill and not be able to move further. When looking at the ball as a wave, imagine that it is a water wave approaching a dam. What happens is that a small amount of water continues past the dam in the form of a wave with a lesser amplitude while the remaining water is stopped by the dam.

In reality, the ability of a particle to perform quantum tunneling is greatly affected by its mass; the smaller a particle is, the more uncertainty it can have in its possible amount of energy, and, therefore, the greater its chances are of overcoming a potential barrier and continuing on the other side. Soccer balls are, actually, too big in mass to be able to roll up a hill without having the minimum energy required—you can slowly roll a million soccer balls toward the hill in the example, and none of them would be able to roll past it by using quantum tunneling. Or, in an example that's possibly clearer to visualize, you could have a million people walk into walls, and none of them would be able to "tunnel" through.

The Source of Life

If quantum tunneling can only occur at the quantum level, with particles too small to be seen by the naked eye, it sounds like an effect that is only relevant to research and technologies that operate at that scale. This couldn't be further from the truth. In fact, there is life on Earth because of quantum tunneling, and that is no exaggeration.

The life cycle starts with the energy we get from the sun in the form of various electromagnetic waves, such as light and infrared radiation. This energy is produced inside the core of the sun through a process called *nuclear fusion*. During nuclear fusion, two atomic nuclei are fused together (as the name implies) to form a larger nucleus, giving off energy in the process. In the sun, the main kind of nuclear fusion that takes place is hydrogen nuclei fusing together to form helium nuclei.

Recall the structure of an atomic nucleus: it consists of protons and neutrons. A hydrogen nucleus is most often simply just a proton, while a helium nucleus has two protons and two neutrons. Protons have a positive electric charge, while neutrons are neutral. What is important to note here is that, like the North poles of two magnets repelling each other, two positive charges naturally repel each other, and fairly strongly at that. The reason they are able to be held together inside most types of nuclei is a particle physics topic of significant depth involving some fundamental forces of nature.

This means that a large amount of energy and some extreme conditions are needed to get hydrogen nuclei to fuse with each other; it is worth it, though, as the energy produced by nuclear fusion is even greater. The core of the sun can provide the necessary conditions to an extent, such as the incredible amount of pressure and temperature required—the inside of a sun is so hot that the particles inside it are in an entirely different state of matter called *plasma*, where electrons are too energetic to stay inside the electron cloud of an atom.

But, even this is not quite enough to get the hydrogen nuclei, which are simply protons, to not just overcome their natural repulsion but to then fuse together afterward in a fairly involved chain of reactions that result in a helium nucleus and a release of fusion energy. Put another way, the hydrogen nuclei inside the sun must overcome an insurmountable potential barrier in order to undergo nuclear fusion—which is where quantum tunneling comes in.

Like the quantum soccer balls slowly rolling toward the steep hill, the protons inside the core of the sun also have wave-particle duality, along with a quantum superposition of several different states. An extremely small fraction of these states (around one in every trillion) have enough energy to overcome the potential barrier, allowing those protons to undergo nuclear fusion and release energy. This is, however, an extremely small fraction of an even larger number of protons to begin with—remember that the sun is more than 100 times larger than the Earth. So, the end result is that a significant number of protons can still undergo nuclear fusion even under suboptimal conditions, causing the release of the energy that eventually reaches Earth, giving life and energy to every living thing on it.

The fusion inside the sun, as well as other stars in our galaxy, is of great practical interest to scientists due to just how much energy is given off by the reaction in proportion to the amount of fuel that goes in. If nuclear fusion can be replicated at a smaller scale on Earth, it could provide a clean, sustainable energy source that is far more productive and efficient than any we currently have. One of the keys to understanding how this might be done is studying the nature of quantum tunneling and potentially figuring out how to manipulate it to our advantage. It is incredibly difficult, not to mention very dangerous, to produce the same conditions on Earth as those found inside the core of the sun; however, since quantum tunneling allows for fusion to take place at less extreme conditions, it could be the key to achieving controlled nuclear fusion at a small scale on the surface of the Earth.

Quantum Tunneling in Practice

The insides of stars are not the only place where quantum tunneling occurs in nature. For example, in the cells of plants, many processes that are necessary for them to function, such as photosynthesis and cellular respiration, involve the movement of electrons and protons across barriers like the membranes (a kind of protective surface) of the cells. These electrons and protons use the quantum tunneling effect to move more efficiently across membranes and other barriers within plant cells.

Even we humans make use of this effect within our bodies. Recent research into the functioning of the brain has shown that its ability to process information and send signals so effectively to the rest of the body is aided, in part, by the quantum tunneling of electrons and other charged particles within the nerve cells in our brain and nervous system (Krämer, 2020). Some of the enzymes in our body, the small proteins responsible for making sure all the chemical reactions that need to take place inside our body do so, also use quantum tunneling in order to perform their function more effectively.

These processes are more efficient for a reason similar to that which makes nuclear fusion in the sun's core more efficient: They allow for certain processes and events to take place without having to put in extra effort or energy to produce more optimal conditions that these processes would otherwise need.

Shifting over to the realm of technology, quantum tunneling is used in the storage of data in flash drives (also known as pen drives or USB drives). Inside these devices are memory cells made up of robust semiconductor materials with thin insulating barriers. Information is stored in and extracted from these memory cells by the controlled movement of electrons through the insulating barriers. The electrons make use of quantum tunneling to make such movements, which is how flash drives can quickly store and deliver information while remaining robust and less likely to be damaged by external causes. More recently, solid-state drives (or SSDs, for short) have made use of a similar internal structure, though at a larger capacity.

The future holds many more possibilities when it comes to making use of quantum tunneling in practical applications. For example, in the fairly new field of nanotechnology (the creation of devices and mechanisms that can perform useful functions at the scale of atoms and particles), the harnessing of quantum tunneling could lead to breakthroughs in anything from computing and communication technology to medical devices. For instance, quantum computers could use quantum tunneling to improve the speed and efficiency of extracting and storing information from their qubits, and medical sensors at a nano-scale could be used to detect the smallest of changes in various biological systems with greater sensitivity and accuracy. There are even more applications to be discovered at the intersection of quantum mechanics with these other fields, from the enhancement of how drugs and medications are delivered to patients to the deeper analysis and manipulation of various states of consciousness in the human body.

SIX

Quantum Entanglement and Non-Locality

Spooky Action at a Distance

In Chapter 1, the Nobel Prize-winning confirmation that the universe is not locally real was treated as a fairly significant and disturbing matter. One of the reasons why the confirmation was so significant was that it provided some of the strongest supporting evidence yet of another peculiarity of quantum physics: *quantum entanglement*. Quantum entanglement, in its simplest terms, is the binding of two or more particles together so that all their properties depend on each other. This might not seem all that strange at first until you take into account the fact that this binding takes effect no matter how far away the particles are from each other or no matter what lies between them.

Consider the following analogy:

Imagine two dancers named Cathy and Dylan, who have been dancing together for years. As a result of this, they share an unusual connection, where their movements and emotions while dancing are completely in sync with one another. When Cathy spins on her toes, so does Dylan. If Dylan suddenly tumbles after a faulty landing, so does Cathy, and for the exact same reason. In essence, Cathy and Dylan are *entangled* with one another. This quantum entanglement comes with a twist, and just the first of many: No matter how far apart they are, their movements will remain in sync. If Cathy twirls in Paris, Dylan will do the same in Tokyo.

You can probably guess why this causes some disruption to the otherwise reasonable idea of locality, which states that objects can only be influenced by their immediate surroundings. If there is a spoon on the table near you, you would only expect it to move if someone picked it up, the table was shifted, or something else in the room had some kind of effect on the spoon (for example, if a large magnet was slowly dragged in). If the spoon were to suddenly start flipping around without anything interacting with it in its locality, you wouldn't be blamed for thinking the spoon was haunted by a poltergeist! These kinds of eerie effects of quantum entanglement are why Albert Einstein famously called the phenomenon "spooky action at a distance" (Muller, 2022).

To look at some of the other peculiarities of quantum entanglement, let's return to the two dancers, Cathy and Dylan. Imagine that Cathy and Dylan are now dancing, each in separate and completely darkened rooms, which means that their movements are now uncertain, even to themselves. They could be performing all manner of routines at any given moment. Given the uncertainty, you would think that they would no longer be in sync; after all, how can Cathy possibly know what Dylan is doing when she is not entirely sure of her own movements, and vice versa? However, this is the second twist of quantum entanglement: even with the added uncertainty, the movements of Cathy and Dylan are still perfectly in sync with each other.

Recall that quantum particles are in a superposition of multiple possible states when not observed and that, because of this, their properties are not entirely known until the particles are observed. However, entangled particles are able to share each other's properties despite the uncertainty inherent to their nature. Another way of viewing this is that each entangled particle has the same probability function for a given property.

Now, imagine the two darkened rooms each dancer is in are on opposite ends of the Earth, and a group of spectators enter each room. When the spectators in Cathy's room switch on the lights to observe her dance, this immediately fixes Dylan's dance into the same set of movements as Cathy's, regardless of whether the spectators in Dylan's room have switched on the lights or not. Since they are on opposite ends of the Earth, this means that Cathy was somehow able to tell Dylan exactly what to do with a speed greater than that of light itself.

The same weirdness would occur if Cathy were to suddenly decide to change her dance to a completely different set of moves. Once again, as she changes her steps, so too does Dylan, even though Cathy can't tell Dylan what to do next or when. Again, she would have had to transmit this information to Dylan at greater than the speed of light.

The speed of light, which is around 30 million kilometers per second, is the ultimate speed limit as far as physics and reality in general are concerned. Only particles that do not have any mass, such as photons, can travel at that speed, and even that only takes place inside a vacuum or a complete absence of any kind of matter whatsoever. Information sent from one point to another needs to travel through some medium and in some form to reach its destination, and it can only do so at speeds equal to or below that of light. The fact that nothing can travel faster than light in a vacuum is considered a fundamental property of the universe.

This, of course, means that what Cathy and Dylan are doing when they instantly sync up with one another sounds impossible. This apparent contradiction of one of the fundamental laws of physics was the subject of a famous paper in 1935 by Einstein together with two other physicists, Boris Podolsky and Nathan Rosen. The initials of their surnames gave rise to the name of the thought experiment they proposed in their paper—the *EPR paradox*. They weren't the only ones to find the results of quantum entanglement contradictory; Erwin Schrödinger also wrote several papers on the topic and coined the term "entanglement" in the first place.

The thought experiment in full involves some in-depth particle physics and mathematical functions, but a simplified version of it is as follows:

Imagine that two particles, A and B, have been entangled with one another. B is kept separate from A and in a space where it is not observed. If the position of A is measured, then, according to entanglement theory, the position of B can be predicted. Then, if the speed of A is measured, the speed of B can also be predicted. This then leads to a situation where both the position and speed of particle B are precisely known without actually measuring either directly. However, this does not line up with the fundamentals of quantum theory, particularly with the uncertainty principle. Therefore, the three physicists argued that quantum theory does not provide a complete picture of what is taking place in this situation. Einstein, in particular, saw it as evidence of his theory of hidden factors that were behind all the quantum peculiarities.

A response to this thought experiment, and one which would pave the way to the 2022 Nobel Prize, came from the physicist John Bell in the 1960s. Bell proposed an equation known as the *Bell Inequality* that could only be solved if the hidden variables Einstein spoke of existed in a quantum system. If the equation could not be solved by results obtained from real-world experiments, then it would be proof that those hidden variables didn't exist and were, therefore, not responsible for the apparent contradictions of quantum entanglement.

This inequality was experimentally proven unsolvable by three Nobel laureates: John Clauser, Alain Aspect, and Anton Zeilinger. By doing so, they confirmed that the principle of locality did not hold true for particles—like the entangled pair of dancers, they could be influenced by entangled particles at any distance. Even the greatest scientists in all of history can be wrong sometimes!

However, one thing to note here is that, contrary to what scientists first believed, the speed of light is not being exceeded. When two particles are entangled, the entanglement becomes an intrinsic behavior to the two of them. This means that when their various properties—their speed, energy, matter wavelength, and so on— become aligned, this takes place without information being transmitted between the two. This is both good and bad: It's good since it aligns with our understanding of physics without causing impossible things to happen. It's bad because it kills the dream that some might have had of faster-than-light communication by way of entangled particles.

Applications of Entanglement

While there are still a lot of mysteries left regarding the nature of quantum entanglement, technology has at least progressed to the point that scientists can entangle particles in the lab with a satisfactory level of success. While photons are the usual particles of choice—a pair of photons that are entangled from the get-go can be created through certain particle collisions—even larger molecules than that of diamond have been successfully entangled together. There are some limitations here as well, though. For example, the particles that are entangled must be created or entangled together after the fact while in close proximity. So, as of now, a quantum object cannot be linked up to one million miles away spontaneously.

However, the fact that we can entangle particles together experimentally has led to a host of possible applications. Although faster-than-light transfer of information is still not quite within our grasp, entanglement can still prove useful in the realm of communication since qubits (quantum bits, the foundation of quantum computing) can be entangled.

In Chapter 3, the method of quantum key distribution, or QKD, was briefly discussed in terms of how it utilizes the uncertainty principle in its operation. To quickly recap, two people can set up a communication channel between themselves that uses qubits to receive and send information. The communication channel uses the properties of quantum particles to ensure it cannot be intruded on without either of the two people noticing. Entanglement plays a role here as well: The qubits in the communication channel can be entangled when the channel is set. This entanglement would then be disrupted if a third party tried to eavesdrop or steal information from the communication channel.

Entangled qubits also allow for improved methods of communication through a method known as *super-dense coding*: Imagine that two people, Alice and Bob, have two quantum systems composed of qubits that can be entangled with each other through external means. Recall that qubits are able to store more data than the typical bit since they are a superposition of several possible states. This means that Alice can send an amount of information that would have needed several classical bits using a much lower number of qubits—the information can be considered to be "super-dense" because of this.

If Alice wants to send information to Bob, they must first entangle their systems together using external means. Then, Alice applies the information to her side of the entangled pair of qubits. When she does this, the qubits on Bob's side will instantly adopt the same properties due to the entanglement. Bob can then access the information by using his system in a similar way to Alice.

Entangled bits can also be useful in quantum computing for performing processes and operations. Since entangled qubits can synchronize their states, this allows multiple sets of qubits to perform the same operations (such as mathematical calculations, storing or extracting information, making amendments, and so on) simultaneously. It is as though, instead of a single horse pulling on a heavy cart, there are multiple entangled horses (quantum entangled, that is, not that their reins are tangled up in any way!) pulling on the same cart, causing it to be shifted much faster. This is what enables quantum computers to perform calculations and other processes much quicker and more effectively than standard computers.

You may have noticed that in the above applications when one set of qubits has its properties set in a certain way, these states are then reflected in the entangled qubits. This transmission of states from one point to another without physically moving the qubits themselves is known as *quantum teleportation*. Unlike the kind of teleportation that makes an appearance in any number of science fiction stories, this does not involve the transportation of physical objects from one point to another—it is the states of the entangled particles that are being instantaneously sent out.

Another application of entanglement is in quantum sensing technology. Since entangled particles take on each other's quantum properties, a group of them can be entangled such that if any one of the particles experiences a disturbance to their state, then all the particles reflect the same disturbance, magnifying the effect of the source of the disturbance. In effect, the entangled particles have become highly sensitive—this is the property used in quantum sensors to detect even the smallest changes in magnetic fields, gravity, and other physical phenomena of interest. The applications of these entangled sensors include medical imaging, mineral exploration under the Earth, and navigation systems that need to react quickly to subtle changes in surroundings.

Entangled Cosmos

It's not possible to directly observe entangled particles out in the far reaches of space. Most of what we can see of space with our current level of technology tends to happen on a far larger scale than the quantum one, after all. But, recent observations have hinted that quantum entanglement might play a part in some of the events seen in outer space (Hawking, 2016).

Take the behavior of particles close to black holes. Black holes, as you might know already, are regions of space with such a strong pull of gravity that not even light can escape them—up to a point.

After all, if a person moves far enough from the surface of the Earth, they will feel the effects of Earth's gravity getting weaker. The same holds true for black holes; at a certain distance away from a black hole's center (this distance is called the *Schwarzschild radius*), light can move past the black hole without being sucked in. Certain interactions with particles at the "edge" of this radius often give out radiation in the form of photons, tiny particles that can escape the gravity of the black hole by moving at very high speed. This radiation, which we receive from the direction of the black hole itself, is called Hawking radiation, named after the famous physicist Stephen Hawking, who theorized such radiation would occur and helped revolutionize our understanding of black holes in the process.

How some particles are formed and emitted from the edge of the black hole is similar to how some particles in the lab can be entangled together. In the case of the black hole, one of the entangled pair escapes the black hole while the other is sucked into its pull —which leads to some interesting possibilities about the nature of black holes and whether we can actually obtain information about their insides by observing the entangled particles released from their vicinities.

Other potential sources of entangled particles and photons in the cosmos include the emissions from supernova explosions (a kind of explosion that stars go through at the end of their lives), the merging of neutron stars (a type of extremely dense star, almost like a black hole but not as gravitationally strong), and even, potentially, in the cosmic microwave background (CMB) radiation, the remaining radiation left over from when the universe first exploded into being.

Beyond the emissions of stars, quantum entanglement also plays a role in the concept of quantum gravity, a theory that tries to bring together the peculiarities of quantum mechanics with the grand scope of general relativity. For example, the amount of entanglement between the various parts of a quantum system can be used to study regions of outer space that have been heavily distorted by strong gravitational fields. Studying entangled particles in space may also reveal insights into how various cosmic structures are formed, from galaxies to planets and everything in between, and in what way these galactic events are influenced by quantum mechanics.

SEVEN

Quantum Computing Basics

Running on Quantum

Imagine that you have been asked to find a book inside one section of a library. For the sake of simplicity, imagine the book exists in that section of the library, so it is simply a matter of finding it. If the books in the library are sorted in alphabetical order, this makes it much easier, of course. You can simply scan the first letters of the book titles until you come across one close to that of the book you're looking for. If the books are not sorted, though, this becomes an immensely difficult task. Without any sorting of the books, you would have to carefully search each shelf until you find the book you want.

Now, imagine that you were not simply yourself but multiple versions of you, all superimposed and entangled with one another. The task of looking for the book still requires the same careful searching of every section, but you now have multiple versions of yourself to simultaneously scan many more shelves at a time than you could on your own. A search that would have taken you hours might now be completed in minutes. This is a glimpse of the potential of quantum computing when it comes to the speed and scale of computations that can be performed.

In classical computers, information is stored and processed in the form of bits. These basic units of information can either be 0 or 1. This stems from the earliest days of computing when data was stored in the form of punch cards. These cards had a grid in which the spaces either had holes or were covered. As our technology advanced, the medium changed from punch cards to many, many other formats, but they all used the same kind of bits for storing information.

In previous chapters, the basics of the quantum bit, or qubit, were touched on. Essentially, qubits can either be 0, 1, or a superposition of these two states with varying probabilities of being either one at a given moment. In addition, qubits can be entangled with one another, which means that they can instantly affect the state of other entangled qubits when they are used in a computation. The combination of quantum superposition and entanglement allows quantum computers to not only make use of more storage capacity per qubit but also use multiple qubits at the same time to perform operations—like the multiple versions of you looking for the book in the library. In terms of computational power, this often gives rise to an exponential increase.

Another reason for the boost in performance is *quantum gates*. In a typical computer, a logic gate is a type of circuit that can perform calculations or *operations* based on the information stored in bits. Logic gates are the building blocks of a computer's processing unit or CPU. Every time an electronic device, such as a computer or smartphone, is processing or doing something with digital information, it uses a network of logic gates to perform the task. Logic gates are also limited by their ability to deal with only 1s and 0s, though quantum gates are not. These are quantum circuits configured to handle the superimposed information stored in qubits, according to the principles of quantum mechanics (Gleiser, 2023).

One of the most fundamental quantum gates is the *Hadamard* gate, which can transform a qubit having either of the base states (so, either a 1 or a 0) into a superimposed state where either 1 or 0 is equally likely. Another important gate is the *Pauli-X* gate, which can flip the state of a qubit (so it can turn a 1 into a 0 and vice versa).

These gates also allow a quantum computer to make use of an effect called *parallelism*. Since multiple qubits can be used at the same time for an operation, this means that a quantum computer can use these qubits to solve a problem by testing out many possible solutions to the problem simultaneously—which, of course, means the solution that works can be reached that much quicker. Going back to the library example, you can reframe the act of finding the book that you're looking for as finding the solution to a problem you're facing (a missing book). Then, having multiple versions of you checking different shelves is similar to trying multiple possible solutions to your problem (since the book could be on any one of the shelves).

Many quantum algorithms (low-level programs that are coded to perform certain types of computational tasks) make use of parallelism in their functioning. One such program is Grover's algorithm, a much quicker version of a typical search algorithm, which allows it to more effectively search even unsorted databases of information (this probably sounds quite familiar at this stage of the chapter!). Shor's algorithm uses parallelism to find all the factors (numbers that can divide a larger number without a remainder) of very large numbers more quickly.

Despite all these benefits of quantum computing, several issues need to be overcome as well at our current level of technology. For example, the type of architecture a quantum computer requires is very different from that of a standard computer and far more demanding in terms of resources and operating conditions. For instance, quantum computers require an extremely low temperature to operate, only a few degrees above *absolute zero* (the lowest possible temperature that can exist). At higher temperatures, the qubits are less likely to maintain the superimposed and entangled states that make them so useful to begin with. To achieve these extremely low temperatures, quantum computers often make use of very large, somewhat impractical, refrigeration units.

Another issue is that qubits can interfere with neighboring qubits within the system. Sometimes, this interference is necessary, such as when qubits are used in tandem for an operation. However, unwanted interference can also happen, causing errors in the information stored in the qubit or some degrading of the data that presents itself as noise. Qubits are also very susceptible to a phenomenon called *decoherence*, in which external influences or disruptions can cause a qubit to lose its superposition and entanglement properties—similar to how observing a quantum state can cause it to lose its superposition and collapse.

The Future of Quantum Computing

Issues aside, the potential of quantum computing has been recognized by multiple tech industry leaders, such as IBM, Microsoft, Google, and Amazon. With the investments and resources they are funneling into their research projects on quantum computing, many avenues are opening up in the development of this technology.

One such area is the building of more robust hardware, making quantum computers easier to scale up in size. At the moment, trying to build one with more than a fairly small number of qubits leads to several complications, from more unwanted interference between the qubits to a significant, impractical increase in all the other connected components, such as wiring and cooling units. Improving this scalability will be necessary to harness more qubits together to solve more complex problems and carry out tasks at a larger scale and with greater efficiency.

To that end, research efforts are underway to increase the number of qubits in a system. The goal is to improve their connectivity with each other and how long a qubit can stay *coherent* in a super-imposed, entangled state. The research also aims to correct and reduce the errors that can form in qubit data due to interference and other factors. New technologies, such as superconducting qubits, trapped ions (a way of using electric fields to support the functioning of qubits), and topological qubits (an arrangement of qubits in a braid-like grid), are being researched as well to over-come the various challenges quantum computers currently face.

Looking ahead, researchers and quantum enthusiasts anticipate that quantum computers will continue to grow in power and capa-bility; by doing so, they will eventually surpass classical computers when it comes to performing certain tasks. As quantum hardware is developed further and becomes more accessible, we can expect to see more widespread use of quantum computing across various industries and eventually, perhaps, even in our homes—thereby mirroring the development path of classic computers.

As quantum algorithms and applications continue to evolve, new use cases and opportunities are likely to be found. From making medicine more personalized to the patient to more accurate modeling of climate changes and even the improvement of supply chain systems, quantum computing has the potential to address some of the most pressing challenges that face society today.

EIGHT

Quantum Cryptography

For Your (Entangled) Eyes Only

The world of spies has far fewer stylish cars and shaken martinis than the movies might imply, and it has far more occasionally tedious busywork such as analyzing reports, monitoring hidden locations, and exchanging secrets through all manner of clandestine means. And, since the information exchanged is usually of the utmost importance at a national level, it needs to be protected from falling into the wrong hands.

The history of encrypting (obscuring information by using a code or key of some kind) goes back centuries and has very little to do with quantum physics for the most part. And even if we knew anything about quantum physics all those centuries ago, it wouldn't have helped much—it was only with the arrival of the computer as we know it in recent decades that quantum mechanics could be thought to have any use in this field. Two main factors contributed to this: The first is that computing allowed for any kind of information, from pictures to sounds to videos, to be stored in the form of 1s and 0s, which could be superimposed on each other as quantum states. The second is that the processing power of computers allowed for far more sophisticated means of obscuring information, such as using complicated mathematical algorithms. The generation and usage of these more complex kinds of encryptions is a task more suited to the capabilities of quantum computers.

Quantum cryptography doesn't need the convoluted mathematical algorithms that many modern encryption systems use, though —not when it has the laws of physics to rely on. In particular, it makes use of the following properties that we have already seen quantum particles and systems exhibiting:

- **Particles are uncertain by nature.** This is due to how particles exist in a superposition of multiple possible positions and states, which means that unless they are observed, the properties of a particle cannot be predicted.
- **Observing or measuring a quantum system alters its properties.** This can also be thought of in another way— any attempt to observe or measure a quantum system will have a noticeable, even measurable, effect on that system.

In addition to the above, photons—like all particles—have a property called *spin*, which is something like the property called *angular momentum* that a macroscopic object gains when it spins. Try thinking of angular momentum as a measure of how quickly something is spinning, as well as how difficult it would be to stop it from spinning. In the case of photons, their spin is binary in nature, meaning that it can be either one of two values. This means that the spin of the photons can be used to denote the 1s and 0s of qubits, much like how the current in a small circuit, being either on or off, corresponds to the 1s and 0s of classical bits.

In Chapters 3 and 6, we took a brief look at quantum cryptography and, in particular, the system known as *quantum key decryption*, or QKD. A common way of simplifying how it works is through the analogy of two people, Alice and Bob, establishing a communication channel with quantum cryptography, with a third party, Eve, trying to hack into it.

If Alice and Bob were spies in the '60s, attempting to give messages to each other through deaddrops and other secretive means, they would always run the risk of a third spy, Eve, being able to intercept their secret messages without either of them being aware. With enough time and effort, Eve could potentially even decode and reveal the messages Alice and Bob were sending each other. Not so in quantum cryptography! In this case, the laws of quantum mechanics themselves act as a shield of sorts against such interceptions.

In QKD systems, information can be sent from Alice to Bob and back through a series of photons being sent across an optic fiber cable. The photons represent the qubits that make up the data being sent and are encrypted by the sender using a special circuit along with polarized filters—which align the positions and orientations of the photons in a particular way—into a specified set of superimposed quantum states. Some QKD systems also include entanglement circuits to entangle the sent photons. The receiver uses its own set of polarized filters and specialized equipment to read the properties of the photons as they are received. When Alice and Bob compare their sets of properties, one such set will match, and this acts as the key the receiver can use to decode the photons and obtain the original information.

The cable doesn't need to be shielded in any way because not only does each photon have its own randomized quantum state, but they are also potentially entangled with one another, making them extra sensitive to outside influences. If Eve tries to eavesdrop on the communication channel, her actions would disrupt the entanglement of the particles and cause decoherence. And, even without the added support from entanglement, Eve's attempts to observe the quantum photons will cause changes to their properties. This change will always be measurable by either Alice or Bob and when they detect it, they will know Eve has been trying to intercept their communications. In this way, QKD systems are effectively immune to being hacked.

That said, QKD systems also have their share of practical challenges to overcome in their current form, mostly issues with setting up a robust infrastructure for the kind of communication channels these systems need. Without these, the photons could degrade over long distances, or their quantum properties could be influenced and even corrupted by environmental factors such as electromagnetic interference.

The polarized filters mentioned above are also the basis of another quantum cryptography system known as *quantum coin-flipping*. Putting aside Eve for a moment, imagine Alice and Bob don't know each other even though they are using the quantum communication channel to share information, and, therefore, they don't entirely trust each other. The orientation of the polarized filters acts as an authentication system, allowing the receiver to verify the information they have obtained is what the sender actually sent.

As an analogy, imagine Bob and Alice are discussing bets on a coin toss through this communication channel, and only Bob has access to the coin. Alice makes a call, Bob tosses the coin, and then sends the information. If Alice calls heads, and Bob says the result is tails, how can she be sure he isn't lying? With quantum coin-flipping, after Alice makes the call, the result of the coin toss itself is encoded as information in the photons Bob sends. If Bob attempted to change the result, this would show as a change in the polarized filter he uses to encode the data. Alice then receives the photons themselves and the orientation of the polarized filter used. To check if the photons have not been altered, Alice first compares the orientation of her filter with that of Bob's—if they line up, she can trust the data. If they don't, meaning Bob attempted to change the result, then she knows Bob isn't being truthful.

Challenges and Frontiers

In the world of cryptography, quantum computers pose a challenge simply by existing. After all, most modern encryption systems make use of complex, tedious mathematical operations (such as needing to find the prime factors of a 100-digit number) that most classical computers cannot hope to solve in a feasible amount of time. But, as you might recall, what takes even super-computers years to do could be accomplished by quantum computers in minutes, including the breaking of these modern encryptions. So, while quantum cryptography systems help some-what by shifting the focus away from these math-based systems, for the many classical systems still in use, there is still a need to develop encryption systems that can withstand an intrusion from a quantum computer.

Just as scalability is an issue with quantum computers, it is also an issue with quantum-based communication systems. As both the distances and amounts of data being handled increase with the size of the communication network, it becomes much harder to reliably manipulate the number of qubits that would need to be used, as well as maintain their coherence over those longer distances and with much more exposure to external factors than they have in the lab. Many of these external factors, from electro-magnetic interference to brute-forced eavesdropping, have a degrading effect on the structure of the quantum system itself.

Most quantum cryptography systems also require specialized equipment and setups (you might have noticed this when QKD was discussed earlier), which are both costly and challenging to construct and maintain. Since quantum tech is still in its early stages, there are very few legal frameworks and operating standards in place that can properly account for how much quantum computers can disrupt our current state of technology. There are issues such as interoperability (how well quantum tech can mesh together with more classical computing systems), security, and ethical and technical concerns that all need to be addressed and worked together on by leading quantum research companies and governments before quantum systems can be more widely used.

Still, it is not all doom and gloom; there *have* been some advance-ments that can help to address some of the bigger issues, such as the resilience of communication systems and the danger of current cryptography systems becoming obsolete.

For instance, there is the ongoing development of *post-quantum cryptography* (PQC), systems designed to withstand attempted hacks from both classical and quantum computers. Researchers are actively looking into new mathematical approaches and types of cryptographic primitives (the basic mechanics a cryptographic system is based on) to develop PQC systems that offer strong security while still being practical for use in the real world.

There is also the field of *homomorphic encryption*, which enables the use of encrypted data without having to decrypt it first. By processing information purely in its secure, encrypted state, homomorphic encryption has a lot of potential to enhance privacy and security, especially in cloud-based computer networks. There have also been promising advancements in increasing the security of *multiparty computation* (MPC) systems, which allow multiple people to connect and work together to perform various computations or tasks without revealing sensitive or confidential information to each other in the process.

NINE

Quantum Biology and Consciousness

From Light to Life

I magine you are enjoying some free time out in a field in summertime, basking in the warm glow of the sun as insects hum in the distance. To capture the happiness and serenity of the moment, you take a picture of yourself surrounded by the greenery of the blades of wild grass by which you're seated, with more trees and bushes in the background. You examine the picture on your camera of choice, assuming it is a digital one, and maybe take a few more pictures until the composition is just right.

In that picture, if you were asked to point out the most significant part of it that involved quantum mechanics, you might point to the particles of light that allowed the picture to exist. After all, how particles of sunlight scatter after reflecting off certain surfaces is very much a quantum phenomenon (as is how the light is captured by the *charge-coupled device*, or CCD, inside a digital camera. These CCDs are a thin wafer of semiconductor material that essentially replaces the film of a traditional camera).

But, within that picture, you would have also captured a large group of living things that make use of quantum mechanics to survive: the blades of wild grass, the trees and bushes, and everything in your surroundings that make use of the process of photosynthesis to capture and store energy from the sun. In Chapter 4, we briefly looked at how photosynthesis makes use of quantum superposition in its functioning, but there is much more to the quantum properties of photosynthesis than that.

A microscopic look at the structure of a leaf, or any other part of a plant in which photosynthesis takes place, reveals the following: There are small cells inside the leaf called *chloroplasts*, which have an outer surface called a *cell membrane*. Embedded in this membrane are "photosystems," which can be thought of as a large group of molecules assembled in one place for the sole purpose of absorbing photons and transferring them to where photosynthesis takes place.

The molecules inside these photosystems are called *chromophores*, which loosely translates from Ancient Greek to "carrier of light." One such type of chromophore is the molecule known as *chlorophyll*, which is what gives plants their distinctive green color. Why green? The simplest way to explain it is that plants evolved to absorb all the possible wavelengths of light given out by the sun in roughly equal amounts. However, the sun gives out more light in yellow and green wavelengths than others, meaning that some of this energy must be reflected to balance things out. It is the light reflected in these colors that we see from plants as a result.

Going back to the chromophores in the photosystems, what they do is absorb photons that strike the surface of the chloroplast and then transfer this energy further into the chloroplast toward what is called the *photosynthesis reaction center*. This center is a specialized group of molecules in which the main chemical reactions of photosynthesis take place. In the process, the energy from the light is used to convert the raw materials for the reaction, such as water and carbon dioxide, into oxygen and the molecule used to store energy in plants known as *adenosine triphosphate*, or ATP. It is from ATP that we humans get energy for our body's various life-supporting processes.

So, where does quantum physics come into all this biology? As it so happens, the transfer of the photon energy from the surface of the cell through the chromophores and into the reaction center takes place at an unusually high efficiency. This could not be explained by a purely classical science perspective, in which the energy basically stumbles its way randomly through the molecules like a pinball being bumped toward the paddles at the bottom or a drunkard staggering through a crowd to find the exit.

However, more recent research has revealed that the process is not quite that random. In fact, the molecules work together in a kind of synchronization as they transfer the energy. In addition, they appear to make use of more energy pathways through the photosystem than can be explained classically (Ball, 2018). How is all of this happening? Quantum superposition explains the multiple possible pathways since it allows for the molecules to exist in varying energy states at the same time.

Since the photons can exist in quantum superposition as well, this effectively amounts to multiple energy transfers within the network of molecules taking place simultaneously, which increases the efficiency of the energy flow. The molecules being in sync with each other can be explained by quantum entanglement:

If the molecules are entangled with one another in a certain pattern, their various states of excitation will allow them to act in harmony as they transfer the energy from the surface to the reaction center. You can think of them as acting similarly to a human chain or a crowd performing "The Wave" at a stadium by coordinating their actions with one another.

The above is an example of quantum mechanics being used in *energy transference* biological processes. Human cells have a similar process for almost the exact opposite: using oxygen to extract energy from where it is stored, producing water, carbon dioxide, and other waste products in the process. This takes place in what are called *mitochondria*, which are tiny subsections found in most of our body's cells. In the mitochondria, instead of photons being transferred, energized electrons move instead, and they make use of quantum tunneling to travel more effectively through the inner membrane of the mitochondria.

Besides the above, the effects of quantum mechanics can be observed in several other biological processes. These can all be categorized into four major groups, one of which we have just discussed: *energy transference*. Another group that was touched on in Chapter 5 is *enzyme catalysis*, which covers all the quantum principles enzymes make use of in order to perform their catalyzing functions.

A catalyst is what allows a chemical reaction to take place with greater efficiency under less extreme conditions. For example, it can cause a large chemical structure to break down into smaller parts without having to increase the temperature or pressure to brute-force the reaction. Enzymes are nature's catalysts and make use of quantum tunneling, in particular, to allow electrons and other particles to move past certain energy barriers without needing an increase in temperature or energy supply.

Sensory processes also make use of quantum physics. Take, for instance, how we use our sense of smell. There are two parts to this process: the cells inside our nose contain complex molecular structures called *olfactory receptors*, which can detect and react to certain chemicals that enter the nose. These olfactory receptors then send a signal to the brain, which translates the reaction into a perception of a characteristic smell. How the receptors detect smell is still being investigated. One quantum interpretation suggests they react to different vibrating frequencies of the incoming molecules in the nose in the same way that the eyes react to different wavelengths of light. Depending on the frequency of vibration and its associated amount of energy, certain electrons at the surface of the receptor get excited and then tunnel through the receptor's molecular structure, which is how the signal is sent out toward the brain.

The last group is *information encoding*, in which quantum mechanics are behind changes in the information stored in certain types of molecules. One example of this is in the theorized mechanism that allows some migrating birds to navigate over long distances by using the magnetic field of the Earth. You can think of this field as lines that certain materials and molecules can align themselves toward—an easy example that comes to mind is a compass needle. These birds have a type of molecule in their eyes called *cryptochrome*, which takes in photons and produces pairs of electrons called *radical pairs*.

In these radical pairs, the electrons can be influenced by the lines of a magnetic field, and like a compass needle, they align themselves in the same direction. When a large enough number of these radical pairs is affected by the magnetic field, the bird is able to detect the alignment of the electrons, and its brain then converts the received signal into a sense of the direction of the magnetic field. The alignment of the radical pairs is boosted by quantum entanglement; when the electrons are entangled, they are more sensitive to the magnetic field outside and can align themselves more quickly with its direction as well.

Quantum Consciousness

Despite our advancements in science and technology, the human mind and consciousness remain mysterious for the most part. Soon, though, that mystery could potentially become slightly easier to investigate with the help of quantum principles and their connection to the essence of consciousness itself.

The theories that scientists have put forward to connect the two are known as *quantum mind hypotheses*, which offer just as many possibilities for unraveling the inner workings of human consciousness as they spark further questions about it. The overall discussion around the idea suggests that consciousness emerges from various quantum processes that take place inside the building blocks of the brain: the brain cells, *neurons*, and the connections between them, the *synapses*.

The classical view of the brain sees it as a complex network of neurons and synapses in which electric currents travel as signals and chemical reactions take place to process or generate them. It's generally thought that consciousness emerges from the interactions of all these smaller parts, like a grand performance that emerges from many people, each playing a small part in it. The quantum mind perspective adds phenomena such as superposition and entanglement to the mix.

In the context of the brain, this could mean that neurons are not only either firing (actively sending or receiving a signal) or passive but can also be in a combination of those states at once. This idea suggests that the brain's processing power could be greatly increased by exploiting this superposition within the neurons. As for entanglement, theorists speculate it could play a role in connecting different regions of the brain to better bring about an emerging consciousness or even in linking individual minds together in a sort of collective consciousness.

However, quantum mind hypotheses are still considered to be highly speculative at present and remain a source of debate within the scientific community. Critics of the various theories argue that the brain is too warm, wet, and noisy an environment for delicate quantum processes to survive intact. You might recall the frigid and well-ordered conditions that qubits in a quantum computer need to function properly; the cells in the brain wouldn't need such extreme conditions to show sustained quantum behavior; however, it would have to be fairly close. Additionally, there is still a lack of universally accepted experimental evidence to support any of these theories.

Despite these challenges, the possible connection between consciousness and quantum mechanics is a captivating field of study. By diving further into the mysteries of the mind and the quantum world in the years to come, we may just uncover new insights into the nature of reality and our place within it—and even find ways to improve the way we perceive and interact with our reality.

Conclusion: Quantum Tech and Future Frontiers

An "Uncertain" Future

In the future, we might all be Schrödinger's Cat, not in the sense that we will be trapped in a box, cursed to be both alive and dead until someone thinks to look for us. Rather, in the sense that the impact of quantum behaviors at the microscopic level could be magnified to the extent that we feel their effects at our macroscopic level of existence.

Imagine a world in which quantum computing is made more accessible for the average person, thanks to advances in superconductors that work at room temperature and materials that allow for qubits to retain their coherence and other quantum properties for extended periods of time. In such a world, jobs that rely on computers to perform tasks could be done in far less time than before, leading to work weeks in tech-heavy industries that consume fewer hours and allow people to be more creative and less drained by their work.

Entertainment could also be boosted in power by these quantum computers. For example, the video games and movies of tomorrow could be delivered through VR and AR with far greater levels of immersion and reactivity, allowing the viewer to truly soak in their experience. And, with the aid of quantum-encrypted communications, they will be able to socialize and network with their friends at a greater speed and with a sense of security.

That sense of security need not be restricted to social media, either! With more widespread adoption of quantum cryptology, everyone from CEOs of major businesses to privacy-conscious celebrities to intelligence agencies of various governments can use these more secure communication channels to send and receive information with less fear of being hacked and a greater ease of mind.

In medicine, doctors could use quantum-powered simulations to figure out difficult diagnoses with more accuracy and efficiency and even be able to fine-tune the treatments to better suit their patients' needs. People could even have diagnostic devices at home that can be used to check their health regularly by a simple insertion of the hand. Aided by the added sensitivity of quantum sensors, these devices could pick up on potential health risks and give the person a chance to prevent unwanted damage to their well-being.

In finance, stock market analysts and investment agencies could make use of a different set of quantum simulations to more accurately predict how and where the market is shifting and use that information to make better financial decisions. In shipping, quantum-powered simulations could map out more efficient routes for all kinds of supply chains, from food to factory equipment. Other fields that rely heavily on the monitoring of different things, from weather stations to navigational systems, can also use quantum sensors to more accurately monitor their environment.

The effects of climate change could be studied in greater detail and perhaps even reversed with the understanding we gain from these studies. In the field of climate and environment monitoring, quantum-powered computers could also allow for more efficient distribution of energy from power plants, leading to a greener and more sustainable future.

Quantum computers could power other devices as well, from AI-driven personal assistants to self-care tools and automated cars. In the world of transportation, self-driving cars could be equipped with quantum sensors and more powerful processors so that they can quickly and safely navigate the roads of tomorrow with greater reliability. This would, in turn, allow them to see more frequent use by people in general, along with a larger presence in public transport. It's not only the cars that could benefit either! Traffic systems and simulations powered by quantum computers could lead to less traffic congestion and less pollution from transport in a city. Trains could even make use of quantum-enhanced levitation to glide smoothly above specialized magnetic tracks, making their travel times drastically lower along with their pollution emissions.

The room-temperature superconductors that can help quantum computers become more mainstream could, themselves, be developed with the help of quantum technology. That's not all: Quantum sensors combined with more accurate microscopes and other observation tech could help scientists study the structure of materials at the atomic level more accurately. This, in turn, can lead to all kinds of exciting new materials to be developed, from better ways of applying medication (such as gels and consumables) to more durable yet comfortable types of clothing, as well as more effective types of batteries and other energy storage devices. Some of these materials could improve renewable energy systems like solar panels and geothermal plants, leading to more countries being able to use these systems for their own power generation needs.

Moving away from the surface of the Earth, quantum technology could have a large impact on both space exploration and the technology used in outer space. Quantum sensors could allow spacecraft navigation systems to be more precise so they can better calculate their paths through space and avoid all manner of dangers, from space debris to meteor fragments. The potential of quantum entanglement to allow quicker communication over long distances means that astronauts and spacecraft can more quickly report back to Earth about their situations and, in turn, receive advice and support that much quicker, too. Telescopes in outer space equipped with quantum sensors and processors could analyze our universe and everything we can see in it with greater depth and detail than ever before. Space, the final frontier, is a great candidate for one of quantum tech's own frontiers.

All of the above may be mostly hypothetical at this stage, but they're not too far off. They are simply a taste of what a quantum-powered world can offer; so many more possibilities could be waiting to be discovered that we can't yet imagine. Of course, with this great power comes great responsibility. As we explore the potential of quantum technology, we need to address the ethical and social issues that come with it.

Societal Impact and Ethical Considerations

A world in which everyone has access to super-fast quantum computers and all the other advancements they enable, seems amazing at first glance. All kinds of major fields, from medicine to transportation, would benefit from this, as discussed above. Computational tasks that used to take years can be done in minutes.

However, in our society, technology doesn't quite work that way. It often takes a while to spread from the more developed countries where it was created and established to the rest of the world. And, given the great leap forward in advancement that quantum technology can give, it would lead to just as large a gap between those who have access to it and those who don't or aren't allowed to. This would become a societal issue that needs serious consideration.

Quantum technologies also bring up all sorts of other tricky ethical questions. For example, consider how easily quantum computers can break through current encryption methods. Until better ones are developed and implemented worldwide, how can we ensure data stays safe in a quantum world? How can privacy, both digital and otherwise, be maintained for the public as a whole? Policies and regulations will need to be put in place to address these concerns, and these policies will need to be able to properly grasp just how disruptive quantum tech can be.

It's not all doom and gloom, though. We have already seen that there are plenty of ways quantum technologies could make the world a better place. As an example, healthcare would benefit from the improved medical sensors and treatment systems that quantum technologies can support. However, even these benefits will bring ethical considerations with them: How would this technology be distributed, especially to those who need it most? Should the technology behind it be open to the public so that everyone can benefit? Would that be allowed by the investors and groups who finance such advancements?

Quantum communication can make our online shopping and networking more secure and much quicker, but in what other ways could that impact society? Could certain jobs and functions of society be made irrelevant or unnecessary by this technology? The same applies to other advancements, such as more reliable automated cars and more capable AI-powered tools.

Ultimately, all these implications of quantum technology are factors that humanity needs to think about, and the sooner we do, the better. For example, we would need to make sure that as many people as possible have access to these advancements and not just a wealthy or elite few; otherwise, the inequality caused by this could cause all kinds of disruptions to our mode of life. Strong ethical guidelines and thoroughly implemented policies would go a long way to ensuring these technologies are used in a way that benefits us all.

If we can find our way through these challenges, the future of quantum technology looks bright. Our world could evolve and improve in ways we can't even imagine. Just as a quantum particle can exist in many possible states at the same time, the possibilities of quantum technology seem to be just as endless. In the end, though, it will be up to us to ensure we harness the power of quantum technologies for the greater good of humanity.

Appendix A: Philosophical Reflections

What Happens to the States in a Quantum Superposition When It Is Observed?

The *Copenhagen interpretation*, which is discussed in this book, suggests that the other states disappear when the quantum system is observed and collapses into a single state as a result. Another view, the *Many-Worlds interpretation*, suggests that at the moment of observation, reality splits into multiple parallel universes (corresponding to each of the possible states that the system can collapse into) that cannot interact with each other. There are many more interpretations that also take into account factors, such as whether the quantum system is itself an observer of its own properties.

Is the Wave Function Describing Reality?

Some scientists believe that the wave function of a particle describes its properties (such as its likely position, its speed, and so on) as they exist in reality, in the same way that Newton's Laws of motion describe the properties of moving things as they are; this is also called the *ontic interpretation*. A conflicting view, the *epistemic interpretation*, states that the wave function doesn't describe reality but is, instead, a mathematical construct that helps us to make sense of a more complex reality. It describes what we're likely to observe in the system when we look at it, but not what it actually is.

Is Reality Objective or Subjective?

Another way of approaching this question is to ask what effect an observer has on the nature of reality. In the Many-Worlds (and Ontic) interpretations, reality is objective since all the possible states that we can observe exist at all times, even if they exist in a parallel state of reality that we cannot access ourselves. In the Copenhagen interpretation and others like it, reality is subjective since it suggests that we change the nature of a system by how we observe it. Put another way, it is our conscious observation of the world that creates the reality we experience, and reality does not have definitive or objective properties of its own.

When Does a Quantum System Collapse?

This might seem an insignificant question at first, but it plays a central role in a spiritual successor to Schrödinger's Cat thought experiment, called *Wigner's Friend* (named after Eugene Wigner, the physicist who devised it). In brief, a scientist, Wigner, has a friend inside a sealed lab, measuring a quantum system, such as a qubit. The friend will get a result of either 0 or 1.

Wigner "opens the box" by asking his friend what the measurement was and gets 0 or 1 as an answer. The issue that arises is whether the system collapsed when Wigner found out the measurement or when his friend took it. Or, did it happen before either of them did anything? Keep in mind that until Wigner asks his friend anything, the contents of the sealed lab are in a super-imposed state to anyone outside of it—including the state of his friend. This issue addresses the role that consciousness plays in an observation.

Appendix B: Quantum Healing Techniques for Self-Healing

Core Tenets

Quantum physics is, in essence, a study of how energy and the particles that possess it behave at a fundamental level. And we, too, are made up of those particles and possess that same energy within us. This energy is not all that follows quantum principles within ourselves. In Chapter 9, the concept of quantum consciousness was discussed, as well as its implications for the structure of our brain and nervous system and how a conscious self can emerge from it.

Quantum healing builds upon the framework that quantum mechanics has given to our understanding of the world and uses it to help you tap into the energies within you, make use of the entangled parts of your body, and harness the power of your consciousness and its connections to the rest of your body, mind, and soul. At its core, quantum healing is based on these core tenets:

- **Our thoughts, emotions, attitudes, and beliefs have a vital impact on our physical well-being and the shaping of our physical reality, and they play an important role in the healing process.** As such, by deepening our understanding of and connection to the above aspects of our consciousness, we can enhance the natural healing mechanisms of our body. This elevated state of consciousness will also attract positive energy into the body, mind, and soul, leading to beneficial outcomes and experiences in daily life.

- **The mind and body are like entangled particles in that they are interconnected and share certain properties.** This connection can be used to treat the root causes of illnesses and mental conditions at a deeper level, leading to a more profound healing and transformation of the soul. A part of this process is cultivating self-awareness and addressing the underlying emotional factors that lead to certain physical ailments and deficiencies in well-being.

- **Quantum healing taps into the body's energy flow and emergent intelligence and makes use of the innate power of our consciousness to enhance healing and improve well-being.** Through practices such as meditation and visualization, individuals can calm their minds, sharpen their focus, strengthen the flow of vital energies within the body, and delve into the deeper levels of consciousness where more profound healing can take place.

The principles of quantum healing, particularly those concerning the flow of energies within the body, can also be seen in other popular concepts in the world of alternative healing, such as the

Chinese concept of the life force, *Qi* (which is a central focus in traditional Chinese medicine as well as in martial arts and ways of life) and the ayurvedic analogous concept of *Prana*, the essence of life; the practice of yoga is a means of stimulating the flow of Prana through the practitioner.

One thing that is important to note is that your emotions are energy, and your body is naturally able to transmute and channel the flow of this energy as long as you do not resist the emotions in any way. Suppressing these emotions can cause imbalances and a blocked flow of energy, which leads to other ailments. Often, simply acknowledging and welcoming your emotions can release any blocked energies within you and stimulate the healing process.

Healing Techniques

Together with fostering emotional awareness, the following techniques can be used to fix any and all potential issues with the flow of energy within your body. It is advisable to try the techniques below only if they resonate with you, as this saves time and energy and moves you along the healing journey more efficiently. Different problems faced by different people require different solutions, and the same holds true here: it's important to pick the techniques that resonate or align with you and your body to better facilitate your healing process. These are some of the most powerful healing modalities that I have personally found effective for myself and my clients.

Mind-body Medicine

This is an umbrella term for a series of techniques that all revolve around one concept: your mind and emotions have a direct influence on your physical health and well-being. These techniques aim to cure or mitigate ailments and other problems in the body by unleashing suppressed or blocked emotions and promoting wellness in the mind. They include, but are not limited to:

- Meditation (to calm and focus the mind)
- Mindfulness (to increase awareness of and attention to the present moment)
- Biofeedback

  ◎ This uses specialized instruments to measure things like heart rate, brain activity, muscle tension, breathing activity, and other indicators of stress and physiological activity. Biofeedback enables people to observe these measurements and consciously alter them to better suit their body's needs.

- Breath Therapy (to reduce stress by slowing and deepening one's breathing)
- Movement therapy

  ◎ These are any physical activities that involve shifting through a certain routine of poses and movements, with the aim of increasing focus and awareness within the body. Examples include yoga, tai chi, dance therapy, and some forms of martial arts.

ThetaHealing

A meditation technique designed by Vianna Stibal that taps into Theta brain waves to bring about physical, spiritual, and psychological healing. When brain cells communicate with each other through electric signals, these signals possess a rhythm that can be detected with an electroencephalograph or EEG. These detected rhythms are known as *brain waves* and are grouped into different categories depending on one's level of consciousness. From the most conscious brain waves to the least conscious, they are Gamma, Beta, Alpha, Theta, and Delta waves.

Theta brain waves are most active when the mind is in between conscious and unconscious states (such as when you are dreaming), where memories and emotions are registered, and attitudes and behaviors are formed. Theta waves are creative and associated with inspiration and spirituality. As such, the ThetaHealing form of meditation that taps into these brain waves allows you to reprogram your mind, clear limiting beliefs, and cultivate positive thoughts.

Emotional Freedom Technique (EFT)

This is a technique by Gary Craig that makes use of *chakras* in the body. To understand what they are, think of your body as a kind of network, with lines of energy connected at various points, in the same way that a fishing net is a series of strings tied together at key points to maintain its structure. The key points where the lines of energy intersect in the body are called *chakras* or *meridian points*. EFT involves a cyclic movement through these chakras to promote the flow of your Qi or Prana.

To perform EFT, first think about a single mental issue that you wish to solve; this can be anxiety, depression, or a traumatic memory. Rank the intensity of the issue on a scale from 0–10, with 10 being the highest intensity possible, and note down the ranking.

While tapping on the outer side of your hand, between your pinky finger and wrist (this is also called the karate chop point and is the part of the hand that strikes the target in a karate chop), think of a simple trigger phrase that acknowledges the issue that you have and conveys that you accept yourself despite it. This can be as simple as, "Even though I [trigger phrase], I choose to love and accept myself." Tapping is done firmly and gently with two or more fingertips. Repeat the phrase you have chosen three times, continuously tapping while you do so.

After this, go through the following sequence of chakras in the body, tapping each point while repeating the trigger phrase chosen above:

1. Directly in the center of the top of your head
2. The beginning of either of your brows, just above and to the side of your nose
3. The bone at the outside corner of either of your eyes
4. The bone under either of your eyes, about an inch below the pupil
5. The point between your nose and upper lip
6. Halfway between the underside of your lower lip and the bottom of your chin
7. Either of the points where your breastbone, collarbone, and first rib intersect, just below the neck
8. At either of the sides of your body, around four inches below the armpit

After this is done, rank the intensity of the issue once more, and then repeat the cycle above. Continue to do so until the intensity is zero or reaches a plateau.

Spiritual Love Affirmations

To increase positivity in your thoughts, hold your left hand over your heart while reading the affirmations below. If you are in a relationship, you can also do these together with your partner by placing your left palm over their heart while they place their left palm over yours.

- My body is designed to self-heal. By changing my beliefs, I will remove any limiting thoughts and unleash all of my body's healing potential.
- I will release all trauma and fear from my life or anyone else's from every cell of my body. I will fill the spaces that remain with divine healing love. I see the world as a wonderful place filled with joy, love, and beauty.
- I am the light, truth, love, and strength of the Creator. I radiate truth so that all darkness, parasites, diseases, and toxins shall pass me by harmlessly.
- I will transform any negative intentions or energies I encounter from any source into a perfectly divine love and forgiveness and then return it to its source.
- I will transform any negative energies within me to that of the highest truth and love and then emit this radiance out into the world.

More Information

To learn more about quantum healing techniques, you can explore my website and social media at the following links:

Website: https://alishakapani.com/
Instagram: @alisha_kapani
Facebook: @ajkapani
YouTube: @alishakapani
TikTok: @alishakapani

Appendix C: Glossary of Key Terms

Algorithm: In computers, an algorithm is a series of instructions in the form of a program or script that allow the computer to perform some kind of task, like a math calculation or trying to find information in a database. Quantum algorithms are designed to be used specifically by quantum computers.

Atom: These are the smallest building blocks of all the chemical elements (such as oxygen, carbon, and hydrogen). They consist of a nucleus at the center, with protons and neutrons inside, and are surrounded by a spherical area called the electron cloud, which contains electrons. When atoms combine to form bigger structures by exchanging those electrons, these bigger structures are called molecules.

Bit: In digital computers and media, information is stored in its most basic form as bits, which are either a 0 or a 1. How this is stored physically depends on the type of device and medium used, but in general, anything that can have two distinct states as a property (which makes it *binary*) can be used to represent a bit. For example, inside modern computing and communications structures, bits are usually represented by whether an electrical current or signal is on or off.

Coherence: In quantum computing, coherence refers to how stable a quantum system is, like a qubit. When a qubit is coherent, it can maintain a superposition of multiple states, including both 0 and 1. In practice, qubits can stay coherent for a certain amount of time (known as *coherence time*) before they go through an interaction with their environment and lose most of their information from the resulting collapse of their multiple states into a single one.

Collapse: An unobserved quantum system has multiple possible states existing at the same time. When this system is observed or disturbed in some way, the system takes on the properties of only one of these states (usually the most likely or probable state). This behavior is what is referred to as a collapse of a quantum superposition.

Decoherence: This is what happens when a coherent quantum state breaks down, usually due to uncontrolled or undesirable interactions with the external environment of the quantum system. During decoherence, the state collapses like it would if it were being observed or measured.

Double-Slit Experiment: This is an experiment that can be used to demonstrate the wave-particle duality of both light and matter particles like electrons. In its simplest form, light or matter is passed through two slits in a barrier and then projected onto a screen on which the resulting pattern can be observed.

Electron: An electron is an elementary particle found inside atoms. Of the three main particles in an atom, it has the smallest mass. It also has a negative electric charge and one of two possible types of spin. Though it is the smallest subatomic particle, it occupies the largest amount of space within an atom.

Electron Microscope: Microscopes use particles as a source of illumination to make out very small details in various objects of study. In electron microscopes, the particles that are used are electrons instead of the typically used light. This allows electron microscopes to examine even smaller details, such as the surfaces of crystal structures or nano-scale objects.

Entanglement: This is a quantum phenomenon in which two particles of the same type can share certain properties, such as speed and spin, even if they are separated by a large distance. These properties are not information that is transferred between the particles but an inherent part of them due to the entanglement.

Information Technology: This is the branch of technology that deals with how information is created, processed, stored, and exchanged between different systems. The kind of systems in question and the information they make use of are generally electronic.

Interference: A behavior of waves that takes place when two or more waves of the same type meet at a point or within an area. When their vibrations build upon each other, it is called *constructive interference*; when their vibrations try to cancel each other out, it is called *destructive interference*.

Lasers: The word "laser" is an acronym for Light Amplification by Stimulated Emission of Radiation. These are beams of light generated by a process called *stimulated emission*. This makes the light travel more coherently and with more focus. Lasers can be much more intense than normal light (enough to cut through solid materials) and can also have specific colors or wavelengths.

Logic Gate: This is a type of electrical circuit in computers that takes in input bits (usually two) and gives out output bits according to how the gate is designed. Some examples include AND gates, which give an output of 1 only when both inputs are 1 (and otherwise give an output of 0), and NOT gates, which invert their inputs (for example, they would change an input of 1 to an output of 0).

Molecule: A molecule is an elemental structure that is made up of two or more atoms grouped and held together by chemical bonds. They are mostly small but can also be fairly large at a microscopic scale. Some examples of this are the molecules for diamond and graphite.

Neutron: One of the particles found inside the nucleus of an atom. They have roughly the same mass as protons and a neutral electric charge. Atoms of the same element can have differing amounts of neutrons, leading to more unstable versions of the element known as *isotopes*.

Photon: When light behaves as a particle, it is, in effect, a discrete amount, or quantum, of energy. In this form, it is referred to as a photon. Photons carry differing amounts of energy depending on the wavelength of their light.

Proton: One of the particles found in the nucleus of an atom. They have roughly the same mass as neutrons but have a positive electrical charge instead. An element is defined by the number of protons inside its nucleus. Although their like charges repel each other, protons are held in a nucleus by a force called the *strong nuclear force.*

Quantization: A quantized property can only have certain discrete values rather than a continuous range of them. As an example, since stairs have steps of specific heights, the elevation of a person on a flight of stairs is discrete, whereas, on the slope of a hill, the elevation would be continuous.

Quantum Communication: This covers any kind of communication channel or network that makes use of the physical properties of quantum mechanics to increase its security and support its functioning. Potential applications of these systems include digital cryptography, enhanced security and privacy, and global networks that are based on satellites.

Quantum Computing: This makes use of the quantum behavior of atoms, molecules, and some types of electric circuits for an atypical, more powerful form of computing. While still in early development, this has the potential to impact all manner of fields, such as security, transport optimization, scientific simulations, material design, and healthcare systems.

Quantum Errors: When a qubit is affected by external factors, the information in it can be corrupted as a result. This is known as a quantum error. Identifying, measuring, and correcting these errors is key to improving the reliability, performance, and scale of operations of a quantum computer.

Quantum Information Science & Technology (QIST): QIST is a multidisciplinary field that combines quantum mechanics with information technology. Research in this field led to the development of various quantum technologies, from quantum computers to quantum sensors.

Quantum Key Distribution (QKD): QKD is a cryptographic system based on the uncertainty principle, in which the parties that are communicating exchange qubits in quantum states instead of electronic signals. QKD uses quantum principles to keep information secure from any kind of attempted intrusion or hack.

Quantum Measurement: Unlike in classical systems, where a measurement simply gives information about the system (such as its mass, length, or temperature), quantum measurements cause the nature of the quantum system to change in the process. The accuracy of certain quantum measurements is incompatible with one another.

Quantum Mechanics: Often interchangeable with quantum physics, this is the study of the nature of existence and our universe at its most fundamental level. It is a radical, indeterministic view of how particles behave at the smallest possible scales of reality and often describes systems as being in a state of uncertainty or flux.

Quantum Sensors: These are sensors that make use of quantum properties, like entanglement, to increase their sensitivity, efficiency, and ability to tell apart different types of stimuli. Potential applications of quantum sensors include early detection and treatment of various medical conditions, improved exploration of the Earth's structure, and the monitoring of biological environments and ecosystems.

Quantum State: The quantum state of a system, such as an electron, is the physical condition in which it exists. Knowing what a quantum state is beforehand allows us to predict the outcomes of experiments performed on that system.

Quantum Superposition: When a quantum system exists in multiple states at the same time, they are said to be in a quantum superposition, in which each possible state has a certain probability attached to it. A state of superposition usually collapses into a single state when observed or measured by an outside party.

Quantum Teleportation: When two particles or quantum systems are entangled, certain properties of one of the particles can be instantly transferred to the other through this entanglement, even if they are a large distance apart. This transfer is known as quantum teleportation and has been performed in labs with structures as large as diamonds.

Quantum Tunneling: This is a wave-like phenomenon in which quantum particles can appear on the other side of an energy barrier at a certain probability, even if the particles in question don't have the energy needed to get past the barrier through classical means.

Qubit: A quantum bit, usually in the form of a quantum particle like an electron or a photon. Like classical bits, these can take on values of 1 or 0. Unlike classical bits, however, a qubit can also exist in a superposition of these two states and even be entangled with other qubits.

Spin: When standard objects such as tops or balls rotate, they have a property called *angular momentum*, which is a measure of how fast they are spinning and how difficult it would be to stop them from doing so. At the quantum level, particles have an analogous property called spin. Spin can only have certain discrete values, such as "up" or "down," which makes it suitable to use as a basis for qubits.

Superconductor: A conductor is a material that can carry an electric current. A superconductor is one in which the current does not experience any resistance as it flows through the material. Most superconductors only gain this property at extremely low temperatures, although research is ongoing into developing materials that offer no resistance to electrical currents—even at room temperature.

Uncertainty Principle: When measuring certain pairs of properties at the quantum level, such as the position and speed of a particle, there is a limit to how accurately these properties can be measured. The more accurately the position is measured, the less accurate the speed will be, and vice versa. In addition, these properties cannot be predicted as they are uncertain unless the quantum system is measured in some way. This is the uncertainty principle.

Wavefunction: This is a mathematical model that describes the quantum state of a quantum system, such as how an electron exists inside an atom. It describes the various characteristics of a particle, such as its position, speed, or spin, as a series of probable values. This is why a wave function is also known as a "probability function."

Wave-Particle Duality: This is the concept that states that quantum matter and light cannot be described as purely a "wave" or a "particle"; they, instead, have some properties in common with waves and others in common with particles. How these properties are exhibited depends on how the matter or light is observed.

Appendix D: Further Reading and Resources

This book was written for people who want to dip their toes into quantum physics. If you would like to go even deeper from here and explore the closely associated field of particle physics along the way as well, the selection below will serve to take you to the next stage of your quantum journey, at the very least.

Historical Perspectives

The Second Creation, by Charles Mann and Robert Crease

A scholarly and journalistic approach to how the study of quantum physics developed, based on extensive interviews with the surviving pioneers of the field as well as their associates. It emphasizes how experiments played a key role in this development, including less commonly discussed topics such as the observation of cosmic rays.

Uncertainty, by David Lindley

This book focuses on how Heisenberg's uncertainty principle came to be and the series of debates between Albert Einstein and Niels Bohr over the philosophical ramifications of quantum theory. It covers the *Old Quantum Theory* period, from when the Bohr model of hydrogen was established in 1913 until quantum mechanics hit its stride in 1927, and nicely shows how Bohr and Einstein often talked past each other in their discussions.

The Age of Entanglement, by Louisa Gilder

This covers the Old Quantum Theory period in a bit more detail but also goes forward into the 1970s, including the first experiments on quantum entanglement. Gilder bases the book around the musings of key figures such as Erwin Schrödinger and Paul Dirac; the content of their letters and other writings is presented as an engaging dialogue.

How the Hippies Saved Physics, by David Kaiser

This is an approachable look at how counterculture and "New Age" thinking played a role in sparking a renewed interest in the foundations of quantum physics in the 1980s, which has now expanded into today's modern research into quantum information. The attempts of the titular hippies to explain ESP through quantum mechanics may not have succeeded, but the analysis of their failures led to some surprisingly useful results.

The Infinity Puzzle, by Frank Close

More of an exploration of particle physics than quantum physics, this details the history of how the Standard Model of particle physics was developed and touches on topics like the development of quantum electrodynamics, the fundamental forces of nature, and what's now called the Higgs mechanism for how particles gain mass. It interestingly describes the many false steps and missed opportunities along the way, giving a good insider look at the process of science as well as core physics.

Further Explorations

The Theory of Almost Everything, by Robert Oerter

This provides a compact overview of the Standard Model of particle physics, which is the collection of particles and forces that make up everything we know about physics at its smallest scale. This is an enormously successful theory that is still only briefly touched upon on the way to more speculative and exotic discussions in most books; this book nicely places it as a center of focus.

QED, by Richard Feynman

This book is based on a series of lectures given by the famous particle physicist Richard Feynman in the early 1980s. This is one of the more comprehensive explanations of the many ideas behind quantum electrodynamics from one of the field's founders. The book exhibits both Feynman's brilliance and his ability to clearly communicate his ideas.

Dance of the Photons, by Anton Zeilinger

A treatment of the physics of entanglement by way of quantum optics. The author is *that* Anton Zeilinger, one of the Nobel Prize winners in 2022, and this is an in-depth look at his field of expertise. It shows the reader how the spookier properties and interactions of entanglement are found by looking at "real" data and various measurements of polarized photons.

Quantum Mechanics: The Theoretical Minimum, by Leonard Susskind and Art Friedman

One of the more math-heavy books on this list, it dives into the details of quantum physics with a greater focus on quantum information than wave-particle duality. Based on a university course Susskind has taught for years, the many equations in the book are presented concisely, and minimal mental effort is required to understand both them and the various ideas they are associated with.

The Quantum Challenge, by George Greenstein and Arthur Zajonc

Another book that makes use of a fair amount of math, in particular that of calculus, which can become overwhelming at first glance. Still, the book presents an overview of noteworthy experiments in quantum physics, especially those that tested the reality of all the more peculiar predictions about the likes of photons and entanglement.

Quantum Enigma: Physics Encounters Consciousness, **by Bruce Rosenblum and Fred Kuttner**

This book heads into the speculative side of quantum theory. One of the less beneficial results of the resurgence of quantum interest in the 1980s was the idea that quantum measurement needs a "conscious observer" to make it work. This unclear, impractical idea has mostly dropped out of real physics, but it also led to some absurd writings about the connection between consciousness and quantum physics; this book is a much more sensible take on the topic in comparison.

Trespassing on Einstein's Lawn, **by Amanda Gefter**

An eclectic and all-encompassing overview of more speculative topics in the field, this book goes beyond the standard low-energy quantum physics to discuss black holes, cosmology, string theory, and more, all encased in a personal narrative involving the author and her father. This book may touch on some heavy ideas but does so with charm and irreverence.

References

Baggott, J. E. (2011). *The quantum story: A history in 40 moments.* Oxford University Press.

Ball, P. (2018, April 10). *Is photosynthesis quantum-ish?* Physics World. https://physicsworld.com/a/is-photosynthesis-quantum-ish

Daniel, S. P., & Daniel, B. (2011). *Quantum techniques: Client manual.* Quantum Techniques, LLC.

Dobrijevic, D. (2022, March 23). *The double-slit experiment: Is light a wave or a particle?* Space. https://www.space.com/double-slit-experiment-light-wave-or-particle

Garisto, D. (2022, October 6). *The universe is not locally real, and the physics Nobel Prize winners proved it.* Scientific American. https://www.scientificamerican.com/article/the-universe-is-not-locally-real-and-the-physics-nobel-prize-winners-proved-it

Gleiser, M. (2023, February 8). *The weirdness of quantum mechanics forces scientists to confront philosophy.* Big Think. https://bigthink.com/13-8/quantum-mechanics-philosophy

H2G2. P. (2006, January 3). *Atoms and atomic structure.* Hitchhiker's Guide to the Galaxy. https://h2g2.com/edited_entry/A6672963

Hawking, S. (2016). *A brief history of time: From the Big Bang to black holes.* Bantam Books. (Original work published 1988)

Krämer, K. (2020, July 30). *Explainer: What is quantum tunneling?* Chemistry World. https://www.chemistryworld.com/news/explainer-what-is-quantum-tunnelling/4012210.article

Lambert, N., Chen, Y.-N., Cheng, Y.-C., Li, C.-M., Chen, G.-Y., & Nori, F. (2012). Quantum biology. *Nature Physics, 9*(1), 10–18. https://doi.org/10.1038/nphys2474

Learn, J. R. (2021, May 5). *Schrödinger's cat experiment and the conundrum that rules modern physics.* Discover Magazine. https://www.discovermagazine.com/the-sciences/schroedingers-cat-experiment-and-the-conundrum-that-rules-modern-physics

McGregor, J. (2023, November 28). *Quantum computing is coming faster than you think.* Forbes. https://www.forbes.com/sites/tiriasresearch/2023/11/28/quantum-computing-is-coming-faster-than-you-think

Muller, A. (2022, October 7). *What is quantum entanglement? A physicist explains Einstein's "spooky action at a distance."* Astronomy Magazine. https://www.astron

omy.com/science/what-is-quantum-entanglement-a-physicist-explains-einsteins-spooky-action-at-a-distance

Orzel, C. (2015, August 15). *Great books for non-physicists who want to understand quantum physics.* Forbes. https://www.forbes.com/sites/chadorzel/2015/08/05/great-books-for-non-physicists-who-want-to-understand-quantum-physics

Orzel, C. (2016, April 18). *How quantum physics starts with your toaster.* Forbes. https://www.forbes.com/sites/chadorzel/2016/04/18/how-quantum-physics-starts-with-your-toaster

Schneider, J., & Smalley, I. (2023, December 1). *What is quantum cryptography?* IBM. https://www.ibm.com/topics/quantum-cryptography

Theta Brain State. (n.d.). Theta Healing. https://www.thetahealing.com/about-theta healing/the-theta-brain-state

Young, H. D. (2012). *College physics.* Pearson Higher Ed.

Quantum Physics for Kids

EXPLORE ATOMS, MOLECULES, & THE MAGIC
OF MATTER WITH FUN ACTIVITIES &
EXPERIMENTS FOR CURIOUS YOUNG MINDS,
AGES 5+

Introduction: Welcome to the Quantum World!

Welcome, curious minds, to a realm of tiny wonders and fantastic discoveries—the amazing world of quantum physics! Have you ever wondered what makes everything around us so magical and exciting? Well, get ready to dive into a universe where the tiniest particles and incredible forces come together to create the world we see!

Imagine a world where things can be in two places at once, where particles talk to each other from far away, and where even a cat can be both asleep and awake at the same time! It might sound a bit puzzling at first, but don't worry—our journey through the quantum world is going to be filled with fun and wonder.

In this book, we'll embark on a fantastic adventure to explore atoms, the building blocks of everything, and how they team up to create magical things called molecules. We'll uncover the secrets of matter and how it can change shapes, from solid ice to melty chocolate, and discover the incredible stories behind some mind-boggling experiments!

But that's not all! We're going to delve into some really cool ideas that might make you scratch your head a bit. We'll talk about a clever cat, some tricky games of hide and seek with particles, and even perform experiments in our minds where things happen in ways we'd never expect!

So, are you ready to join us on this incredible adventure into the mysterious and magical world of quantum physics? Get your imagination ready, because we're about to discover some of the most amazing things our universe has to offer!

Get set, quantum adventurers! Our journey is about to begin!

ONE

What's an Atom?

Imagine a tiny building block that makes everything!

Welcome, curious minds! Today, we're going to embark on an incredible journey into the heart of the smallest things in our universe—atoms! Imagine that everything you see, touch, or even imagine is made up of these tiny, super-duper small building blocks called atoms.

Let's take a moment to think about a world where everything is made up of these little building blocks. Picture your favorite toys, your pets, even the air we breathe—all made of countless numbers of these tiny, invisible particles!

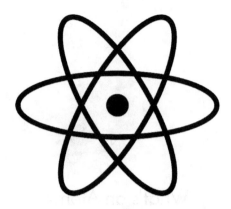

Atoms as Building Blocks

So, what exactly are atoms? Well, atoms are like the building blocks of a giant LEGO castle or the tiny beads in your favorite bracelet. They're so small that you can't see them with your eyes, but they're incredibly important because they make up everything around us!

Imagine atoms as super tiny balls. These balls are so small that if you lined up a billion of them, they'd only be about as wide as the dot on this page! They're the basic units that make up everything in the whole wide world.

Atoms are like LEGO blocks or marbles!

Think of atoms as those magical LEGO blocks you love to build with. Each atom is like a single LEGO block, and when you put many atoms together, just like when you connect many LEGO blocks, they create something new and amazing!

Or, imagine atoms as tiny marbles. When you collect lots of marbles and put them together in different ways, they can create all sorts of things—just like how atoms come together to create everything around us!

Activity: Let's Build an Atom!

Are you ready for some fun? Let's create our own atom using colorful playdough or by drawing one! Imagine an atom has a round center called the nucleus. Let's make a small ball for the nucleus. Around the nucleus, add some tiny balls or sticks to represent the electrons moving around it. Voila! You've created your very own atom!

TWO

Inside the Atom Adventure!

Let's shrink and explore inside an atom!

Welcome back, explorers! Last time, we discovered that everything around us is made up of tiny building blocks called atoms. Now, it's time to journey inside these amazing atoms and find out what's hiding within!

Introduction to Protons, Neutrons, and Electrons

Imagine atoms as tiny solar systems! At the center of our atom-world are these special, super-tiny parts called protons and neutrons. These are like the stars and planets, and they're snuggled together in a place we call the nucleus.

And zooming around the nucleus, like speedy comets, are even tinier particles called electrons. They zip around the nucleus in special paths called orbits.

Playful Visualization

Think of the nucleus as a cozy group of friends huddling together in the middle of a big playground. The protons and neutrons stick close together, like the best pals they are, while the electrons whizz around them, playing a never-ending game of tag!

Activity: Balloon and Confetti Party!

Are you ready for a bit of fun with electricity? Grab a balloon and rub it on your hair or a cozy sweater. Now, hold the balloon close to some tiny bits of paper or confetti. Watch what happens! The confetti will stick to the balloon, just like electrons stick to atoms because of something called electric charge!

Wow, wasn't it amazing to see how the balloon could attract the confetti? That's a bit like how electrons and atoms play together!

Next time, we'll dive into how these playful atoms team up to create fantastic things called molecules!

THREE

The Magic of Molecules!

Imagine atoms holding hands and making something new!

W elcome back, curious minds! Last time, we ventured inside atoms and met their tiny inhabitants—protons, neutrons, and electrons. Today, get ready to witness the fantastic team-up of these atoms as they create something truly magical —molecules!

Picture molecules as teams of atoms working together, just like a squad of friends joining hands to create something amazing! When atoms hold hands—or more accurately, share their electrons—they form these wonderful groups called molecules.

Examples of Familiar Molecules

Think about the air you breathe or the water you drink—these are all made up of molecules! For instance, water (H_2O) is a fantastic team of two hydrogen atoms and one oxygen atom. Even the air we breathe, which is mostly made of oxygen molecules (O_2), is created by teams of oxygen atoms working together.

Activity: Mix and See!

Let's have some colorful fun! Take two glasses of water—one with red food coloring and the other with blue. Carefully pour them into a third glass. What do you see? The colors mix and swirl together, just like how atoms come together to create new colors in molecules!

Wow, isn't it amazing how atoms join forces to create these marvelous molecules? Next time, we'll explore how these incredible molecules play a big role in creating different forms of matter —solid, liquid, and gas!

FOUR

The Wonderful World of Matter!

Imagine matter as a shape-shifter!

Welcome back, explorers! We've uncovered the secret teams of atoms forming fantastic molecules. Now, let's delve deeper into the enchanting world of matter—where things can change shape, size, and even how they feel!

Definition of Matter: Anything That Takes Up Space and Has Mass

Think of matter as the stuff that fills up space and feels different to touch. Whether it's your favorite toy, a soft pillow, or the air you breeze through, it's all matter! Anything that takes up space and has weight is made of matter.

Solid, Liquid, and Gas States

Matter can transform into three main forms: solids, liquids, and gases. Imagine a solid as something you can hold, like a rock or your toys. A liquid, like water, can flow and take the shape of its container. And gas, like the air we breathe, can fill any space it's given!

Activity: Matter Changing Its Form!

Let's have some hands-on fun with matter! Take an ice cube and watch it melt in your hand or put it under the sun. What happens? The ice changes from solid to liquid! You can also melt chocolate and then let it cool back into a solid. That's matter changing its form!

See how matter can change from solid to liquid to gas? Isn't that fascinating? In the next chapter, we'll dive into some super exciting ideas about cats, uncertainty, and some really curious experiments that might leave you scratching your head!

FIVE

Schrödinger's Curious Cat: An Unusual Tale

Once upon a time, in a house filled with cozy nooks and sunlit spots, there lived a special cat named Schrödinger. Now, Schrödinger wasn't your ordinary feline—he was an extraordinary cat of paradoxes, where naps and adventures danced together in a most peculiar way.

Paws and Possibilities

Picture a fluffy cat nestled in a sunny corner, peacefully snoozing away with his eyes closed tight. But wait, something curious happens! While Schrödinger dreamt of chasing cosmic mice or taking grand leaps among twinkling stars, he was also exploring the house, chasing yarn and playing hide-and-seek with invisible friends. How can he be asleep and awake all at once? It's a puzzle of playful possibilities!

Dreamy Adventures Unravel

As the sunbeams painted patterns on the floor, Schrödinger's dreams and his wide-eyed escapades began to blur. In his dreams, he'd soar through quantum realms, hopping between alternate universes, chasing rainbows made of stardust. Meanwhile, in reality, he'd scamper around, paws batting at imaginary wonders—two adventures intertwined into one!

Purr-fect Paradox Unleashed

Now, imagine Schrödinger's human friends—they couldn't quite understand their furry companion's peculiar behavior. "Is he asleep?" they'd wonder. "Or is he wide awake?" It was a delightful puzzle for everyone in the house, a whimsical mystery wrapped in a purring, fuzzy bundle of fun!

Tails of Sleepy-Wakefulness

In Schrödinger's world, a doze and a frolic could twirl together, making his adventures a magical blend of dreams and reality. He'd nap, whiskers twitching to imaginary tales, while simultaneously galloping across the house in playful pursuit of invisible wonders. It was as if his dreams tiptoed into his wakeful moments, creating a joyous, puzzling dance of sleepiness and alertness!

Schrödinger's Whisker-twirling Tale

In the enchanting tale of Schrödinger's curious cat life, dreams and wakefulness swirled together like colors on an artist's palette. His playful paradox became a charming mystery, leaving everyone with giggles and grins, marveling at the delightful confusion of a cat both asleep and wide awake. And as the sun set, Schrödinger curled up, whiskers twitching to tales unknown, ready to embark on another whimsical journey of paradoxical adventures in dreamland and reality!

Activity: Cat's Nap Hide-and-Seek

Materials Needed:

- Stuffed toy cat or a picture of a cat
- A small box or a toy bed for the cat
- A soft blanket or a cloth to cover the cat

Activity Steps:

1. **Meet Schrödinger's Cat:** Introduce the child to the stuffed toy cat or show them a picture of a cute cat.

Explain that this cat is special and sometimes does something very surprising!

2. **Cat's Sleeping Spot:** Show the child the small box or toy bed and place the cat inside it. Explain that the cat loves to take naps and sometimes likes to hide in its cozy spot.

3. **Cat's Mystery Nap:** Cover the box or bed with the soft blanket or cloth, making it a cozy hiding place for the cat. Explain that while the cat is hidden, it's taking a nap, but it might also be awake at the same time!

4. **Guess the Cat's State:** Ask the child to guess whether the cat is napping or playing while it's hidden. Encourage them to imagine what the cat might be doing inside without peeking.

5. **Reveal the Surprise:** After a short while, remove the cloth and show the cat in its hiding spot. Discuss with the child how they guessed the cat's state while it was hidden. Emphasize that until they saw the cat, it was both napping and awake in a playful way.

6. **Talk About the Surprise:** Explain that just like how it was tricky to know if the cat was napping or awake while hidden, in a story about a special cat named Schrödinger's cat, it's like the cat is doing two things at once until someone looks at it.

7. **Embrace the Surprise:** Encourage the child to appreciate surprises and that sometimes things can be a bit surprising or tricky to guess, just like the playful mystery of Schrödinger's cat!

SIX

Exploring Uncertainty and Duality: A Playful Puzzle Adventure

H ey there, curious minds! Let's dive into a world where surprises reign supreme—where toys play hide-and-seek and guessing games turn into playful mysteries. Welcome to the whimsical playground of uncertainty and duality, where things aren't always what they seem!

Guessing Games Galore

Imagine having a super fun guessing game where things aren't where you'd expect them to be. Picture this: you close your eyes, and your toys vanish into thin air! But wait, they're not gone—just hiding in plain sight. That's a bit like uncertainty, where things aren't predictable, and surprises hide around every corner!

Waves, Particles, and Magical Transformations

Now, let's talk about a playful idea—imagine if your favorite ball could be both bouncy like a ball and flowy like a river at the same time. That's a bit like what happens in the quantum world! Things there can behave like both waves and tiny, zoomy particles, just like your toys sometimes feel like they have magical powers!

Mystery in the Quantum Realm

In the wibbly-wobbly, topsy-turvy world of quantum physics, uncertainty and duality are like playing pretend with your toys. Sometimes, they act super surprising, making you scratch your head in wonder! It's a bit like when you guess where your toy is, and it's not where you thought—but that's the fun part of the game!

Playful Surprises and Puzzling Possibilities

In the magical land of uncertainty and duality, surprises twirl around like your favorite spinning top. Things don't always follow the rules you expect, creating a jigsaw puzzle of giggles and excitement! But that's what makes it an adventure—a chance to explore and giggle at the mysterious and playful side of the quantum world!

Mysteries to Giggle About

So, my playful pals, uncertainty and duality in the quantum world are like secrets waiting to be uncovered. They're puzzles with giggles hidden inside—surprises that make you go, "Wow!" Remember, it's okay not to know everything because in this magical realm, mysteries are like little treasures, waiting for curious explorers like you to discover them!

Activity: Surprise Box

Materials Needed:

- A small box or container with a lid
- Various small toys or objects that fit inside the box (e.g., small plush toys, toy cars, building blocks)

Activity Steps:

1. **Prepare the Surprise Box:** Gather the small toys or objects and place them inside the box. Make sure the child doesn't see what's inside. Close the lid securely.
2. **Shake and Guess:** Hand the closed box to the child and ask them to guess what might be inside without opening it. Encourage them to use their imagination and guess based on the sounds or feelings when they shake the box.
3. **Unveil the Surprise:** After guessing, open the box to reveal the contents. Discuss with the child if their guesses were correct or if they were surprised by what was inside. Talk about how it's sometimes tricky to predict what's inside a closed box just by shaking or guessing.
4. **Discussion Time:** Explain in simple terms that sometimes surprises or unpredictability can be fun, just like not knowing what's in the box until it's opened. Relate this idea to the uncertainty principle, mentioning that in the quantum world, there are surprises or things we can't predict about tiny particles.

5. **Embrace Surprises:** Encourage the child to appreciate surprises and unexpected things in everyday life, just like the surprise of discovering what's inside the box. Emphasize that it's okay not to know everything and that surprises can be exciting!

SEVEN

Unraveling the Double-Slit Mystery: A Fun Experiment Adventure

Hey there, curious minds! Imagine a magical game where things change just by peeking—welcome to the exciting world of the double-slit experiment! This experiment is like a playful puzzle that shows us how sneaky particles behave when they think no one's watching. Let's dive into this curious adventure together!

The Mysterious Game Begins

Imagine setting up a special game—let's call it the Double-Slit Game! You invite tiny, zoomy particles to play, but here's the twist: they act super tricky! When no one's looking, they zoom through two tiny holes, like they're saying, "Catch us if you can!" But when you peek, they switch their game plan—oh, what a puzzling play!

The Sneaky Particle Dance

In this wibbly-wobbly experiment, particles act like secret agents, doing a surprise dance when no one's watching! They zoom through the holes and create a cool pattern on the other side, almost like they're painting with tiny paint brushes. But when someone tries to peek, they shy away, changing their dance steps. It's a playful particle dance-off!

Peek-a-Boo, Particles!

Let's play a game of peek-a-boo with these sneaky particles! When you try to catch them in the act, they change their moves, creating a whole new pattern on the wall. It's as if they're saying, "Aha! You can't see us now!" Their playful dance changes whenever someone tries to sneak a peek, leaving behind a trail of giggles and mystery.

Fun with Tiny Surprises

In the super cool world of the double-slit experiment, particles keep surprising us like playful little friends! Their mischievous ways show us that they can act differently when we're not looking. It's like having a secret that only particles know—making this experiment a delightful game full of twists and turns!

Playful Particle Puzzles

So, my playful pals, the double-slit experiment is like a magical game where particles do the silliest dances and create secret patterns. It's a puzzling mystery that teaches us that particles can be a bit shy when they know we're peeking. Remember, even the tiniest particles love a good game of hide-and-seek!

Activity: Wavy Water

Materials Needed:

- A shallow, rectangular container (like a baking pan or a plastic tray)
- Water
- Two thin barriers (these can be pieces of cardboard or paper)
- A flashlight or a small, bright light source (like a phone's flashlight)

Activity Steps:

1. **Set Up the Experiment:** Fill the shallow container with water, about halfway. Place it on a stable surface.
2. **Create the Barriers:** Take the pieces of cardboard or paper and place them vertically in the water, leaving a small gap between them. Ensure the barriers are close enough so that they create two openings or slits in the water.
3. **Let the Light Shine:** Position the flashlight or bright light source to shine light towards the barriers in the water. Adjust the angle so that the light passes through the gaps between the barriers.
4. **Observe the Patterns:** Turn off the room lights or darken the area around the container. Look at the patterns created by the light passing through the gaps. You should see alternating bands or patterns of light and shadow on the surface opposite the light source.
5. **Experiment with Waves:** Now, gently move the barriers closer together or farther apart. Observe how the patterns

on the surface change as you adjust the gaps between the barriers. Notice how the light behaves differently based on the width of the gaps.

6. **Discussion Time:** The experiment with the water and light shows a bit like the double-slit experiment in the quantum world. Just like how the light created patterns when passing through the barriers, particles also create patterns when passing through slits, showing both wave-like and particle-like behavior!

7. **Encourage Exploration:** Play with the barriers and observe how changing the width affects the patterns. Notice how the light behaves differently depending on the setup.

EIGHT

Quantum Superposition: The Wondrous Trick of Being Everywhere!

Hey, curious minds! Imagine if you could be in two places at the same time—sounds like a magical adventure, right? Well, welcome to the incredible world of quantum superposition, where things can do the most amazing disappearing-reappearing acts! Let's dive into this mind-bending idea together using a playful analogy that'll make it super easy to understand!

The Marvelous Vanishing Act

Imagine you're a magician, and you have a special trick up your sleeve—let's call it the Disappearing-Reappearing Magic! You step into a magical box and say the magic words: "Presto, change-o!" Poof! You vanish! But wait, here's the twist: you also appear on the other side of the room at the same time! That's a bit like what happens in the quantum world!

The Sneaky Quantum Magic

In the amazing world of quantum physics, tiny particles do the most incredible tricks! They can be in two places at once, just like your magical disappearing-reappearing act. Picture this: a particle here and also there, doing a playful dance of "I'm here, I'm there, I'm everywhere!" It's a giggle-inducing magic show of particles!

Juggling Two Places at Once

Now, imagine you're juggling two balls at the same time—one in each hand. That's a bit like what particles do in the quantum world! They can juggle being in two places simultaneously, just like you juggle those balls. It's a super-duper balancing act of particles that'll leave you scratching your head in wonder!

Superposition: The Quantum Showstopper

In the enchanting realm of quantum superposition, particles put on the grandest show! They can be like two friends at two different parties, having fun at both places at the same time. It's like having a twin who can be in two places, doing two different things simultaneously—it's that mind-blowing!

Quantum Magic Unleashed!

So, my fantastic friends, quantum superposition is like having a magic wand that makes things appear in two places at once! It's a playful trick that particles do in the quantum world, leaving us in awe and wonder. Remember, even in the quantum world, the coolest magic shows are full of surprises!

Activity: Magic Juggling Act

Materials Needed:

- Three different-colored balls or toys (e.g., small plush toys, balls, or building blocks)

Activity Steps:

1. **Gather the Toys:** Collect three different-colored toys or balls. Make sure they're easy to hold and toss around.
2. **Pick a Spot:** Find a comfortable space where you can move around freely without any breakable items nearby.
3. **The Juggling Game:** Start by holding two toys in one hand and the third in the other hand.
4. **Juggle Like Magic:** Begin tossing the toys gently in the air. But here's the fun part—instead of catching them, pretend they're all in the air at once! Move your hands to mimic catching the toys, but don't actually catch them. Keep them 'suspended' in the air using playful gestures.
5. **Counting the Toys:** Ask the child how many toys they see in the air. They might say three, even though you're pretending to hold them all at once.

6. **Reveal the Magic:** Explain that just like the toys seemed to be in two places at once during the juggling game, in the quantum world, tiny particles can act in a similar way —being in two places or states simultaneously!

7. **Repeat and Explore:** Encourage the child to try the 'magic juggling' act themselves. Let them pretend to suspend all the toys in the air at once and see how many they can visualize.

NINE

Quantum Entanglement: The Incredible Twin Connection of Particles!

Hey there, curious explorers! Let's dive into a mind-boggling world where particles do the most amazing magic trick of all time—quantum entanglement! Imagine having a secret connection, just like how friends share secret messages. Well, in the quantum world, particles do the same—it's like having a special twin connection that's super-duper cool!

The Twin Telepathy Game

Imagine you and a friend have a magical connection, like secret twin telepathy! Your friend goes far, far away, to the moon, maybe, while you stay on Earth. Now, here's the fantastic part: when you think of a special word, your friend instantly knows it! That's a bit like how particles share secrets in the quantum world—they're connected no matter how far apart!

Spooky Actions at a Distance

In the enchanting land of quantum physics, particles have a super mysterious bond called entanglement. It's like having a pair of magical socks—if you pull one, the other one moves, even if they're in different drawers! When something happens to one particle, the other reacts instantly, as if they're sharing a secret handshake across galaxies. It's a bit spooky, but oh-so-fascinating!

The Quantum Teleportation Dance

Now, picture this: particles do a playful dance where one does a spin, and the other twirls at the same time—even if they're galaxies apart! It's like having a synchronized dance party where moves match instantly. That's the magical world of entangled particles—a dance of mysterious connection that's absolutely out of this world!

The Twin Connection's Mystery

In the fantastical world of quantum entanglement, particles act like best buddies who know each other's secrets. Even if they're on opposite ends of the universe, they talk without using words—it's a secret language of particles! Their connection is like having a special phone line that works without wires, instantly passing messages like a lightning bolt.

Quantum Twin Secrets Revealed!

So, my amazing pals, quantum entanglement is like having a secret twin language between particles, making them the best friends of the quantum world. They share secrets instantly, no matter how far apart they are, creating a magical connection that'll leave you grinning in wonder! Remember, even particles have best friend secrets!

Activity: Quantum Twin Dance

Materials Needed:

- Two colorful ribbons or strings of equal length (different colors or patterns work well)
- A partner or a friend

Activity Steps:

1. **Get Set:** Choose a partner or a friend to do this activity with you. Each person holds one end of a ribbon or string, making sure they are equal in length.

2. **Twirl and Dance:** Stand a few steps away from each other. Now, start moving around while holding onto your ends of the ribbons. As you move, twirl and dance around in different directions, making playful moves with the ribbons.

3. **Mirroring Moves:** Try to mirror each other's movements using the ribbons. If one person twirls the ribbon, the other person should try to do the same. It's like doing a dance routine, but with ribbons!

4. **Connected Moves:** Here's the fun part! Even when you're both far apart, try moving in a way that the ribbons seem to respond to each other's movements. For example, if one person makes a big loop with their ribbon, the other person should try to do something similar at the same time.

5. **Observing the Connection:** After dancing and twirling with the ribbons, take a moment to notice how your moves seem connected. Discuss with your friend how, even though you're holding separate ends of the ribbons, your movements affected each other's actions!

6. **The Entangled Ribbons:** Explain that this activity shows a bit like how entangled particles behave. Just like how you and your friend's moves with the ribbons seemed connected, entangled particles act in sync with each other, no matter how far apart they are!

Conclusion: Our Quantum Journey!

Congratulations, fellow adventurers! We've completed an incredible journey into the fascinating world of quantum physics. Together, we've unraveled the mysteries of atoms, peeked inside their tiny worlds, and discovered the magic of molecules and the ever-changing nature of matter. But our journey doesn't end here —it's just the beginning of our quest for knowledge and wonder!

We've embarked on a whirlwind tour through some mind-bending ideas in the quantum realm—Schrödinger's playful cat, the puzzling principles of uncertainty and duality, the curious double-slit experiment, and the fantastic phenomena of superposition and entanglement. These ideas might seem a bit puzzling, but they're the keys to understanding our amazing universe!

Remember, the quantum world is full of surprises and mysteries waiting to be explored. It's okay if some ideas still seem a little tricky—science is all about asking questions and seeking answers!

As we wrap up this adventure, keep your curious spirit alive. Keep asking questions, keep exploring, and never stop wondering about the world around you. Who knows, maybe someday you'll uncover even more incredible secrets hidden in the quantum realm!

So, fellow adventurers, until our next journey into the wonders of science, keep imagining, keep exploring, and keep embracing the magic of discovery. The quantum world awaits your curious minds!

Farewell for now, quantum explorers!

Bonus Experiments: Uncover the Magic of Quantum Physics!

Experiment 1: Colorful Light Show with a Prism

Step 1: Gather your magical prism and a flashlight!

Get ready to uncover the magic of light that's connected to our quantum journey! Find a prism—a special crystal—and a flashlight.

Step 2: Let the light shine through!

Hold the prism in the sunlight or shine the flashlight through it. This magical crystal splits the light into colors, just like how we discovered in our quantum adventure that light behaves both as waves and particles.

Step 3: Watch the magic unfold!

The light passes through the prism and transforms into a beautiful, colorful rainbow on the wall! Remember how we learned about the duality of light in our quantum journey? This experiment shows how light behaves like both waves and tiny particles called photons.

Experiment 2: Magnetic Exploration Adventure

Step 1: Gather your magnetic treasures!

Get ready to uncover the mysterious forces that connect to quantum physics! Find magnets—those cool things that attract and repel each other.

Step 2: Let's play with attraction and repulsion!

Watch how the magnets stick together or push away. It's a bit like quantum entanglement where particles can be connected, even when far apart! These invisible forces show us a bit about the spooky actions happening between particles in the quantum world.

Experiment 3: Sound Waves Jam Session

Step 1: Join the music band—grab your instruments!

Get ready to explore vibrations, just like particles in the quantum world! Pick instruments or use your magical voice.

Step 2: Let's make some noise!

Play your instruments or sing your favorite song. Feel the vibrations—just like how particles vibrate in the quantum world. Listen closely to the magical world of sound waves, much like how we explored waves in our quantum journey.

Experiment 4: Coin Flipping Fun & Probability Play

Step 1: Get your lucky coins ready!

Let's dive into probability—the chance of something happening. Find shiny coins—these relate to the uncertainty principle in quantum physics!

Step 2: Flip, flip, hooray!

Flip the coin and see if it lands heads or tails. Discuss the chances, much like how quantum physics deals with probabilities and uncertain outcomes in the quantum realm.

Experiment 5: Shadow Play & Light Magic Show

Step 1: Gather your magical light source!

Time for shadow fun that connects to our quantum exploration! Find a light source like a lamp or flashlight.

Step 2: Time for shadow magic!

Shine the light on objects and watch shadows appear. Explore how light interacts with objects—just like how we learned about light interacting with particles in our quantum journey, creating mysterious effects like the double-slit experiment.

References

OpenAI. (2023). Conversations with ChatGPT. Retrieved [Jan 9, 2024], from https://www.openai.com/chatgpt/

www.ingramcontent.com/pod-product-compliance
Lightning Source LLC
Chambersburg PA
CBHW071248050326
40690CB00011B/2304